Collins

The Shanghai Maths Project

For the English National Curriculum

一课一练

Practice Book 4A

Series Editor: Professor Lianghuo Fan

UK Curriculum Consultant: Paul Broadbent

Collins

William Collins' dream of knowledge for all began with the publication of his first book in 1819.

A self-educated mill worker, he not only enriched millions of lives, but also founded a flourishing publishing house. Today, staying true to this spirit, Collins books are packed with inspiration, innovation and practical expertise. They place you at the centre of a world of possibility and give you exactly what you need to explore it.

Collins. Freedom to teach.

Published by Collins
An imprint of HarperCollins*Publishers*
The News Building
1 London Bridge Street
London
SE1 9GF

HarperCollins *Publishers*
1st Floor
Watermarque Building
Ringsend Road
Dublin 4
Ireland

Browse the complete Collins catalogue at
www.collins.co.uk

MIX
Paper from responsible sources
FSC™ C007454

www.fsc.org

This book is produced from independently certified FSC paper to ensure responsible forest management.

For more information visit:
www.harpercollins.co.uk/green

The Shanghai Maths Project (for the English National Curriculum) is a collaborative effort between HarperCollins, East China Normal University Press Ltd. and Professor Lianghuo Fan and his team. Based on the latest edition of the award-winning series of learning resource books, *One Lesson, One Exercise*, by East China Normal University Press Ltd. in Chinese, the series of Practice Books is published by HarperCollins after adaptation following the English National Curriculum.

Practice Book Year 4A has been translated and developed by Professor Lianghuo Fan with the assistance of Ellen Chen, Ming Ni, Huiping Xu and Dr Jane Hui-Chuan Li, with Paul Broadbent as UK Curriculum Consultant.

© HarperCollins*Publishers* Limited 2018

© Professor Lianghuo Fan 2018

© East China Normal University Press Ltd. 2018

10 9 8 7 6 5 4

ISBN 978-0-00-822613-8

Translated by Professor Lianghuo Fan, adapted by Professor Lianghuo Fan.

British Library Cataloguing in Publication Data

A catalogue record for this publication is available from the British Library.

Series Editor: Professor Lianghuo Fan
UK Curriculum Consultant: Paul Broadbent
Publishing Manager: Fiona McGlade
In-house Editor: Mike Appleton
In-house Editorial Assistant: August Stevens
Project Manager: Emily Hooton
Copy Editors: Tanya Solomons, Tracy Thomas and Karen Williams
Proofreader: Joan Miller
Cover design: Kevin Robbins and East China Normal University Press Ltd.
Cover artwork: Daniela Geremia
Internal design: 2Hoots Publishing Services Ltd
Typesetting: 2Hoots Publishing Services Ltd
Illustrations: QBS
Production: Sarah Burke
Printed and Bound in the UK using 100% Renewable Electricity at CPI Group (UK) Ltd

Contents

Chapter 1 Revising and improving

1.1 Warm-up revision

 Learning objective Use strategies to calculate mentally and using written methods

 Basic questions

1 Calculate.

(a) 200 + 300 = ☐

(b) 460 − 230 = ☐

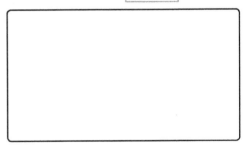

(c) 170 + 330 = ☐

(d) 1000 − 400 = ☐

(e) 470 − 230 = ☐

(f) 660 − 600 = ☐

(g) 500 + 500 = ☐

(h) 480 – 230 = ☐

(i) 290 + 110 = ☐

2 Find patterns and fill in the boxes.

(a) 148 + 152 = ☐

138 + 162 = ☐

128 + 172 = ☐

☐ + ☐ = ☐

(b) 855 – 170 = ☐

865 – 170 = ☐

☐ – 170 = ☐

885 – 170 = ☐

(c) 220 + 348 = ☐

220 + 350 = ☐

220 + 352 = ☐

☐ + ☐ = ☐

3 Complete the facts.

(a) A square has _____ sides and _____ angles.

The lengths of all the sides are _____ .

All the angles are _____ angles and equal to _____ degrees.

(b) A cuboid has _____ vertices, _____ faces and _____ edges.

(c) A cube has _____ faces.

The shape of each face is a _____ .

4 Fill in the answers.

(a) Write a suitable unit in each space.

(i) Erin's weight is about 32 _____ .

(ii) There are 24 _____ in a day.

(iii) Tom spent 10 _____ eating a piece

of bread that weighed about 100 _____ .

(b) Write a suitable number in each space.

(i) 1000 grams = ☐ kilogram(s)

(ii) 1 hour 17 minutes = ☐ minutes

(iii) 2 minutes = ☐ second(s)

(iv) 35 g + 423 g = ☐ g

(v) 2012 was a leap year and it had ☐ weeks and ☐ days,

or in total ☐ days.

5 Use the column method to calculate.

(a) 362 + 607 = ☐

(b) 1000 − 872 = ☐

(c) 631 − 89 + 452 = ☐

6 Write a suitable number in each ☐.

(a)
```
    7   8   ☐
+   ☐   2   6
─────────────
    9   ☐   4
```

(b)
```
    ☐   9   3
−   3   ☐   4
─────────────
    2   0   ☐
```

Challenge and extension questions

7 Think carefully and then fill in the boxes with suitable numbers.

(a) $18 \times 9 + 18$

$$= 18 \times \boxed{}$$

$$= \boxed{}$$

(b) $21 \times 6 - 6$

$$= \boxed{} \times \boxed{}$$

$$= \boxed{}$$

8 Make a count. In the figure on the right, there are $\boxed{}$ letter 'A's and $\boxed{}$ triangles.

1.2 Multiplication tables up to 12 × 12

Learning objective Recall multiplication facts and division facts

Basic questions

1 Write the multiplication facts and complete the table.

×	1	2	3	4	5	6	7	8	9	10	11	12
10	10				50					100		
11		22				66					121	
12			36				84					144

2 Use the multiplication table above to complete two multiplication sentences with the given product and write the multiplication fact. The first one has been done for you.

Product: 60	Product: 88
5 × 12 = 60 (or 6 × 10 = 60)	
12 × 5 = 60 (or 10 × 6 = 60)	
Multiplication fact:	Multiplication fact:
5 × 12 = 60 (or 6 × 10 = 60)	

Product: 110	Product: 132
Multiplication fact:	Multiplication fact:

3 Fill in the boxes to make the equations true.

(a) $96 = 8 \times \boxed{} = 12 \times \boxed{} = 6 \times \boxed{}$

(b) $44 = 4 \times \boxed{} = 11 \times \boxed{} = 2 \times \boxed{}$

(c) $108 = 9 \times \boxed{} = 12 \times \boxed{} = 3 \times \boxed{}$

4 Calculate.

(a) $12 \times 5 = \boxed{}$ (b) $11 \times 6 = \boxed{}$ (c) $9 \times 11 = \boxed{}$

(d) $24 \div 2 = \boxed{}$ (e) $5 \times 11 = \boxed{}$ (f) $7 \times 11 = \boxed{}$

(g) $12 \times 12 = \boxed{}$ (h) $33 \div 11 = \boxed{}$ (i) $11 \times 10 = \boxed{}$

(j) $100 \div 10 = \boxed{}$ (k) $11 = \boxed{} \div 12$ (l) $12 = \boxed{} \div 6$

5 George collected 121 stickers and gave them all to 11 friends equally. How many stickers did each friend get?

Answer: _____

6 Meena reads 12 pages of a book every day. The book has 308 pages. Meena has been reading the book for 11 days.
How many pages has she read? How many pages are left?

Answer: _____

Challenge and extension questions

7 A deck of cards can be counted in 11s and in 12s respectively, without any left over.

What is the smallest possible number of the cards in the deck?

Answer: _____

8 All pupils in Year 4 are grouped equally for a school activity. If each group consists of 10 pupils, there are 6 pupils left ungrouped. If each group consists of 11 pupils, there are 4 pupils left ungrouped. Given that the number of pupils in Year 4 is fewer than 150, how many pupils are there in Year 4?

Answer: _____

1.3 Multiplication and division (1)

Learning objective Multiply and divide 3-digit numbers by a single-digit number

Basic questions

1 Use the column method to calculate. Check the answers to the questions marked with *.

(a) $89 \times 6 =$ [　　]

(b) *$801 \div 9 =$ [　　]

(c) $8 \times 356 =$ [　　]

(d) $293 \times 5 =$ [　　]

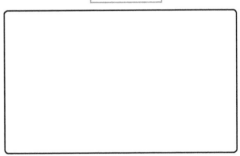

(e) *$769 \div 7 =$ [　　]

(f) $464 \div 8 =$ [　　]

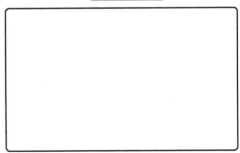

2 Write >, < or = in each ◯.

(a) 87 ÷ 3 ◯ 78 ÷ 3

(b) 150 ÷ 5 ◯ 105 ÷ 5

(c) 264 ÷ 4 ◯ 272 ÷ 4

(d) 75 ÷ 5 ◯ 75 ÷ 3

(e) 392 ÷ 2 ◯ 392 ÷ 8

(f) 756 ÷ 8 ◯ 756 ÷ 7

(g) 96 ÷ 8 ◯ 84 ÷ 7

(h) 650 ÷ 5 ◯ 990 ÷ 9

(i) 610 ÷ 5 ◯ 738 ÷ 6

3 Work these out step by step.

(a) 6 × 187 − 216 =

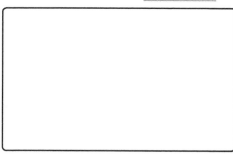

(b) 585 ÷ 5 + 78 =

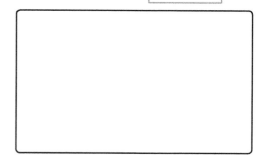

(c) 408 × 7 + 973 =

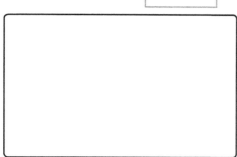

4 Jack is 8 years old. He and his younger brother, Joe, went to a theme park with their parents on a Sunday. The admission ticket was £36 for an adult and £25 for a child under 12. How much did they pay for the admission tickets in total?

Answer: _____

5 The table below shows information about three pupils in a running exercise.

	Shia	Yee	Mo
Time (minutes)	5	8	4
Distance (metres)	675	768	544

Who ran the fastest? _____

Who ran the slowest? _____

 Challenge and extension questions

6 Yasmin went to buy some pens with £54. There were three prices for the pens: £9, £6 and £18 each. Help Yasmin to work out different combinations of pens so the total cost is exactly £54 each time.

Answer: _____

7 Fill in the boxes with suitable numbers to make the calculations true.

(a)

```
    □ 8 □
  ×     □
  _____
  7 0 □ 5
```

(b)

```
         □ □   4
     3 ) □ □   2
         6
       _____
         □ □
         □     1
       _____
           1   2
         □ □
       _____
             0
```

1.4 Multiplication and division (2)

Learning objective Multiply and divide 3-digit numbers by a single-digit number

Basic questions

1 Use the column method to calculate.

(a) $130 \times 6 = $ ☐

(b) $620 \div 4 = $ ☐

(c) $560 \div 8 = $ ☐

(d) $537 \div 9 = $ ☐

2 Work out each calculation step by step.

(a) 462 × 7 + 367 = ⬚ (b) 780 × 5 − 1000 = ⬚

(c) 658 + 326 − 558 = ⬚

3 Write the number sentences and then work out the answer.

(a) What number times 8 is 1000?

Number sentence: _____

Answer: _____

(b) 7 times a number is 721. Find this number.

Number sentence: _____

Answer: _____

(c) What is 5 times 465?

Number sentence: _____

Answer: _____

(d) The divisor is 65, the quotient is 8 and the remainder is 9. What is the dividend?

Number sentence: _____

Answer: _____

4 Application problems.

(a) A box of cola has 10 bottles. Each bottle has 380 ml. Louis has drunk 4 bottles. How many millilitres has he drunk? How many millilitres of cola are still left in the box?

Answer: _____

(b) Jaya's father is 40 years old this year. This is exactly 4 times Jaya's age. How many years younger than her father is Jaya?

Answer: _____

(c) One lap of the running track in a school sports field is 200 metres long. Katy has run 4 laps and Sajid has run 1 kilometre.

(1 kilometre = 1000 metres or 1 km = 1000 m)

(i) How many metres has Katy run? _____

(ii) How many laps has Sajid run? _____

(iii) How many more metres has Sajid run than Katy? _____

 Challenge and extension questions

5 A school bought some footballs and basketballs. There are 20 more basketballs than footballs. The number of basketballs is 3 times the number of footballs. How many footballs did the school buy? How many basketballs?

Answer: _____

6 In the following column calculation, what numbers do A, B, C and D stand for in order to make the calculation correct?

$$
\begin{array}{r}
A\ B\ C \\
\times \qquad C \\
\hline
D\ B\ C \\
\hline
\end{array}
$$

A = ☐   B = ☐   C = ☐   D = ☐

1.5 Problem solving (1)

Learning objective Use strategies to solve multiplication and division problems

Basic questions

1 The school library has 48 science books. The number of storybooks is 15 more than 3 times the number of science books. How many storybooks does the library have? Use the line model to help work out the answer.

48

3 × ?   15

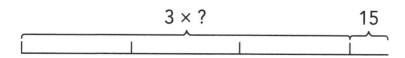

Number sentence: _____

Answer: _____

2 Tom is 11 years old. His father is 5 years younger than 4 times Tom's age. How old is Tom's father? Draw a line model to analyse.

Number sentence: _____

Answer: _____

3 There are 34 star-shaped balloons. The number of moon-shaped balloons is 3 times the number of the star-shaped balloons.

(a) How many balloons are there altogether?

Method 1:	Method 2:

(b) How many more moon-shaped balloons are there than star-shaped balloons?

Method 1:	Method 2:

4 Finn, Kayo and Aliyah all like collecting stamps. Finn has collected 24 stamps. He has 8 stamps fewer than Kayo. The number of stamps that Aliyah has collected is 3 times the number of stamps that Kayo has collected. How many stamps does Aliyah have?

Answer: _____

Challenge and extension questions

5 There are 360 ash trees in a woodland. There are 10 more oak trees than twice the number of ash trees. The number of birch trees is 10 fewer than twice the number of ash trees.

(a) How many oak and birch trees are there in total?

Answer: _____

(b) How many fewer birch trees are there than oak trees?

Answer: _____

6 865 trees are planted along the side of a lake. Two maple trees are planted between every two willow trees. How many willow trees and how many maple trees are planted?

Answer: _____

1.6 Problem solving (2)

Learning objective Use strategies to solve multiplication and division problems

Basic questions

1 Look at the diagrams below, write the number sentences and calculate.

(a)

Black sheep ⎰ 32 ⎱

White sheep

⎱ How many sheep are there in total?

Number sentence: _____

Answer: _____

(b)

Pens ⎰ 19 ⎱

Pencils

How many?

Number sentence: _____

Answer: _____

2 Write the number sentences and then find the answers.

(a) A is 357. B is 3 times as great as A. What is the sum of A and B?

Number sentence: _____

Answer: _____

(b) A is 357. It is 3 times as great as C. What is the sum of A and C?

Number sentence: _____

Answer: _____

3 3000 kilograms of cement were delivered to a warehouse in the morning. The amount of cement delivered in the afternoon was 2000 kilograms more than twice the amount delivered in the morning. How many kilograms of cement were delivered in the afternoon?

Answer: _____

4 A high-speed train travels 240 kilometres in one hour. This is 4 times as fast as a car. How many more kilometres does the high-speed train travel than the car in one hour?

Answer: _____

5 A farm has 120 cows. The number of sheep is 3 times the number of cows. The number of chickens is 5 times the number of cows.

(a) How many fewer cows than chickens are there on the farm?

Answer: _____

(b) How many more chickens than sheep are there on the farm?

Answer: _____

(c) How many cows, chickens and sheep are there altogether?

Answer: _____

6 There were 72 pupils in the library and in the PE hall altogether. After 12 pupils left the hall and entered the library, there were 3 times as many pupils in the library as in the hall. How many pupils were there in the library and in the PE hall at first?

Answer: _____

7 A 90-metre-long rope is cut into two parts. The first part is 2 metres shorter than 3 times the length of the second part. How long is the second part?

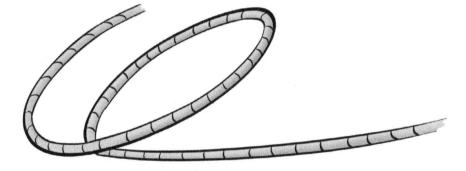

Answer: _____

1.7 Fractions

Learning objective Recognise, find and calculate fractions of shapes and quantities

Basic questions

1 Fill in the blanks with fractions.

(a) The cake was divided into 2 equal parts.

Each part is ☐ of the cake.

(b) The chocolate bar was divided into 8 equal parts.

Each part is ☐ of the chocolate bar.

(c) The square is divided into 4 equal parts.

Each part is ☐ of the whole.

2 Write a fraction to represent the shaded part of each figure.

(a) 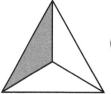 (b) 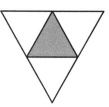 (c) 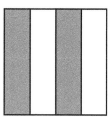 (d)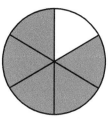

☐ ☐ ☐ ☐

3 Shade each figure below to represent the fraction given.

(a) $\frac{1}{4}$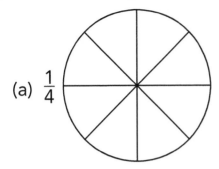

(b) $\frac{1}{4}$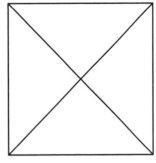

(c) $\frac{2}{9}$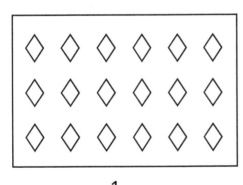

(d) $\frac{5}{6}$

4 Circle the objects in each diagram to show the fraction given below.

(a)

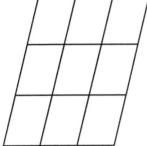

$\frac{1}{6}$

(b)

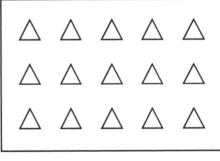

$\frac{2}{5}$

5 Fill in the boxes.

(a) $\frac{1}{2}$ of 20 ▲ is ☐ ▲.

(b) $\frac{1}{5}$ of 20 ▲ is ☐ ▲.

(c) $\frac{3}{4}$ of 20 ▲ is ☐ ▲.

Challenge and extension question

6 Use fractions to represent the shaded parts in the figures below.

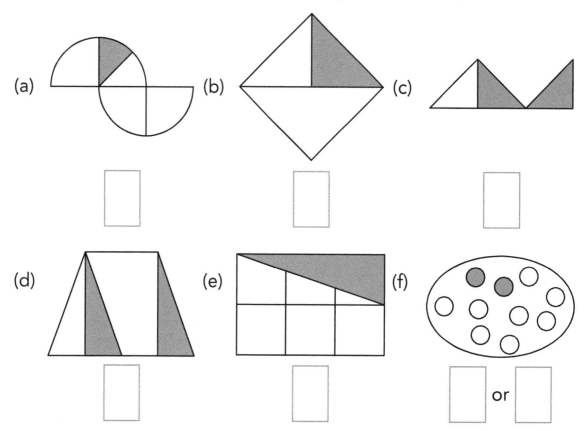

Chapter 1 test

1 Calculate mentally and then write the answers.

(a) 150 + 450 = ☐

(b) 680 − 320 = ☐

(c) 540 − 450 = ☐

(d) 12 × 12 = ☐

(e) 64 ÷ 8 ÷ 4 = ☐

(f) 9 × ☐ = 27 + 9

(g) 200 ÷ 50 = ☐

(h) 300 ÷ 30 = ☐

(i) 11 × 12 = ☐

(j) 70 + ☐ = 120

(k) 720 ÷ ☐ = 9

(l) ☐ × 8 = 80

2 Use the column method to calculate. Check the answers to the questions marked with *.

(a) 1000 − 888 = ☐

(b) 987 − 789 + 123 = ☐

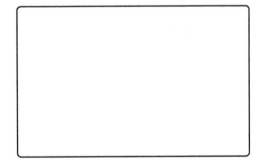

(c) 3 × 285 = ☐

(d) 720 ÷ 4 = ☐

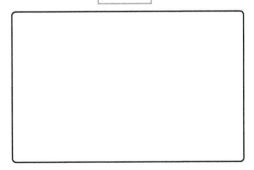

(e) *89 ÷ 7 = ☐

(f) *365 ÷ 9 = ☐

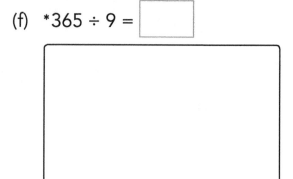

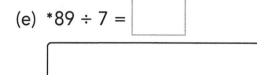

3 Work these out step by step. (Calculate smartly when possible.)

(a) 472 − 148 − 152 = ☐

(b) 213 − 43 + 87 = ☐

(c) 751 + 273 − 451 = ☐

(d) 25 × 78 × 4 = ☐

(e) 1008 ÷ 7 ÷ 9 = ☐

(f) 24 × 5 + 333 = ☐

4 Fill in the spaces.

(a) **1** **2** **3** **4** **5** **6** **7** **8** **9**

Use these numbers to form two 3-digit numbers. Do not use the same digit twice. The greatest possible difference between two such numbers is ☐.

(b) (i) $500 \div \boxed{} = 10$ (ii) $\boxed{} \times 7 = 4900$

(iii) $\boxed{} \div 8 = 30$

(c) Write >, < or = in each ◯.

(i) $48 \div 4 \bigcirc 84 \div 4$ (ii) $219 \div 3 \bigcirc 192 \div 3$

(iii) $606 \div 6 \bigcirc 630 \div 6$ (iv) $360 \div 6 \bigcirc 360 \div 60$

(v) $615 \div 3 \bigcirc 615 \div 5$ (vi) $428 \div 4 \bigcirc 428 \div 2$

(d) Write a suitable unit in each space.

(i) Tom is 120 _____ tall.

(ii) The mass of an egg is 70 _____.

(iii) The height of a desk is 100 _____.

(iv) The mass of a watermelon is 4 _____.

(v) Tower Bridge in London is 244 _____ long.

(vi) A football costs 15 _____.

(vii) Joe's height is 1 _____ and 45 _____.

(viii) A car travels at 60 _____ per hour.

(e) A circle was divided into 6 equal parts.

Each part is ▢ of the circle.

(f) 7 lots of $\dfrac{1}{▢}$ is equal to seven ninths.

(g) The denominator of a fraction is 34 and the numerator is 11 less than the denominator.

The fraction is ▢.

(h) There are ▢ lots of $\dfrac{1}{21}$ in $\dfrac{3}{21}$.

(i) A 1-metre-long ribbon was divided into 5 equal pieces.

The length of each piece is _____.

4 of these piece put together are _____ metres long.

5 Shade each figure below to represent the fraction given.

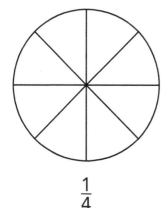

$\dfrac{1}{4}$

$\dfrac{5}{9}$

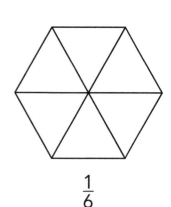

$\dfrac{1}{6}$

6 What fraction of each figure is shaded? Write two fractions for each.

 ▢ or ▢ 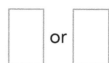 ▢ or ▢  ▢ or ▢

7 Application problems.

(a) Leon cut a 15-metre-long ribbon into lengths of 3 metres each. How many cuts did Leon make?

Answer: _____

(b) Trees are planted every 4 metres along the side of a road. Kiran ran from the first tree and stopped at the tenth tree. How many metres did she run?

Answer: _____

(c) 42 pots of flowers are placed along both sides of a path. They are placed 3 metres apart. How long is the path?

Answer: _____

(d) When Jay started learning to type, he could type 15 words in one minute. Now he can type 7 words more than 4 times the number of words he to type at the beginning. How many words can he type in one minute now?

Answer: _____

(e) In a food wholesalers, a box of onions costs £18 and a crate of tomatoes costs £28. The price of rice is £42 per sack. Toni and Rosa run a restaurant. They went shopping with £400.

   (i) They first bought 2 crates of tomatoes and 3 sacks of rice. How much did they pay?

     Answer: _____

   (ii) They then bought 3 boxes of onions. How much did they pay for the onions?

     Answer: _____

   (iii) How much money did they have left?

     Answer: _____

Chapter 2 Numbers to and beyond 1000 and their calculation

2.1 Knowing numbers beyond 1000 (1)

Learning objective Place value of 4-digit and 5-digit numbers

Basic questions

1 Calculate mentally and then write the answers.

(a) 63 + 91 + 20 = ☐

(b) 21 − 3 + 6 = ☐

(c) 54 − 46 − 2 = ☐

(d) 9 × 1 × 4 = ☐

(e) 8 × 2 × 7 = ☐

(f) 60 ÷ 4 ÷ 5 = ☐

(g) 120 ÷ 5 ÷ 6 = ☐

(h) 0 × 3 × 0 = ☐

(i) 70 ÷ 10 × 9 = ☐

(j) 180 ÷ 10 ÷ 3 = ☐

(k) 98 + 32 + 44 = ☐

(l) 30 ÷ 10 × 9 = ☐

2 Read and write the numbers given below.
The first one has been done for you.

(a) The longest road distance in Britain is from Land's End to John O'Groats: 1407 km.

Read as: <u>one thousand four hundred and seven</u>

1407 = ☐ 1000 ☐ + ☐ 400 ☐ + ☐ 0 ☐ + ☐ 7 ☐

(b) The highest mountain in Britain is Ben Nevis. Its peak is 1344 m above sea level.

Read as: _____

1344 = [ ] + [ ] + [ ] + [ ]

(c) The flight distance from London to Sydney, Australia, is 16 983 km.

Read as: _____

16 983 = [ ] + [ ] + [ ] + [ ] + [ ]

(d) The Great Wall of China is about 6700 km long.

Read as: _____

6700 = [ ] + [ ] + [ ] + [ ]

(e) The Nile in the north-east of Africa is the longest river in the world. It is 6853 km long.

Read as: _____

6853 = [ ] + [ ] + [ ] + [ ]

3 Complete the place value charts.

(a) 1001

Ten Thousands	Thousands	Hundreds	Tens	Ones

(b) 9212

Ten Thousands	Thousands	Hundreds	Tens	Ones

(c) 7035

Ten Thousands	Thousands	Hundreds	Tens	Ones

(d) 10 535

Ten Thousands	Thousands	Hundreds	Tens	Ones

4 Fill in the spaces.

(a) There are [ ] thousands and [ ] hundreds in four thousand.

(b) 5 tens make [ ]. 5 hundreds make [ ].

5 thousands make [ ].

(c) From the right, the first place of a number is the _____ place. The place to its left is the _____ place.

(d) From the right, the fourth place of a number is the

_____ place. The place to its right is the

_____ place. The place to its left is the

_____ place.

(e) The number 4075 consists of 4 _____,

_____ tens and 5 _____.

It reads _____.

The difference between this number and the greatest 3-digit

number is _____.

5 Multiple choice questions. (For each question, choose the correct answer and write the letter in the box.)

(a) 8 is in both the thousands place and the hundreds place.

This number is ☐.

    A. 88 000   **B.** 8800   **C.** 880

(b) In a number, the ten thousands place is 1, the thousands place is 2 and the hundreds place is 3. The numbers is ☐.

    A. 123   **B.** 1230   **C.** 12 300

(c) In a 4-digit number, the sum of all the digits is 36.

The number is ☐.

    A. 3600   **B.** 9999   **C.** 6666

(d) In a 4-digit number, the product of all the digits is 1.

The numbers is ☐.

    A. 1000   **B.** 1100   **C.** 1111

Challenge and extension questions

6 In a 4-digit number, the sum of all the 4 digits is 2.
Write all such numbers.

_____

7 Give some real-life examples in which numbers more than 1000 are used.

2.2 Knowing numbers beyond 1000 (2)

Learning objective Compare and order 4-digit numbers

Basic questions

1 Count in multiples and fill in the boxes.

(a) 1, 2, 3, ☐, ☐, ☐, 7, ☐, ☐, ☐.

(b) 10, 20, 30, ☐, ☐, ☐, 70, ☐, ☐, ☐.

(c) 100, 200, 300, ☐, ☐, ☐, 700, ☐, ☐, ☐.

(d) 1000, 2000, 3000, ☐, ☐, ☐, 7000, ☐, ☐, ☐.

(e) 0, 25, 50, 75, ☐, ☐, ☐, 175, ☐, ☐, ☐.

2 Follow the instructions to write the correct numbers.

(a) Numbers that come before and after each number.

(i) ☐, 1738, ☐   (ii) ☐, 9999, ☐

(iii) ☐, 4106, ☐   (iv) ☐, 6000, ☐

(b) Whole tens that come before and after each number.

(i) ☐, 3900, ☐   (ii) ☐, 1550, ☐

(iii) ☐, 7809, ☐   (iv) ☐, 6000, ☐

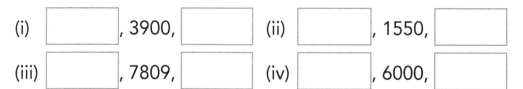

(c) Whole hundreds that come before and after each number.

(i) [ ] , 2657, [ ]   (ii) [ ] , 4505, [ ]

(iii) [ ] , 7790, [ ]   (iv) [ ] , 6000, [ ]

(d) Whole thousands that come before and after each number.

(i) [ ] , 1067, [ ]   (ii) [ ] , 9222, [ ]

(iii) [ ] , 4050, [ ]   (iv) [ ] , 6000, [ ]

3 Count and complete the number patterns.

(a) 234, 235, 236, [ ] , [ ]

(b) 1000, 1025, 1050, [ ] , [ ] , 1125

(c) 500, 2500, 4500, [ ] , [ ]

(d) 9000, 8000, 7000, [ ] , [ ] , 4000

(e) 3500, 3000, 2500, [ ] , [ ] , [ ] , [ ] ,
[ ]

4 Use these numbers to complete each sentence.

| 10, | 920, | 6000, | 5010, | 1000, | 3280, | 8880, | 6540, | 4990 |

(a) The numbers greater than 3000 but less than 6000 are

_____ .

(b) The number that is 5000 more than 1000 is [ ] .

(c) The whole thousands are _____ .

(d) Put all the above 4-digit numbers in order starting from the greatest:

[ ] [ ] [ ] [ ] [ ] [ ] [ ] .

Challenge and extension questions

5 Choose from the digits to make the following numbers:

| 2, 3, 5, 6, 8, 0 |

   (a) the smallest possible 4-digit number

   (b) the smallest possible 5-digit number

   (c) the greatest possible 4-digit number

   (d) the greatest possible 5-digit number

6 When you write numbers from 7000 to 8000, you need to write

7 _____ times, 8 _____ times and

9 _____ times.

2.3 Rounding numbers to the nearest 10, 100 and 1000

Learning objective Round numbers to the nearest 10, 100 and 1000

Basic questions

1 Write the number that each letter stands for.

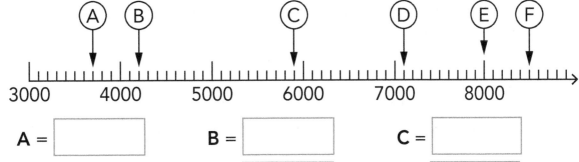

A = ☐   B = ☐   C = ☐

D = ☐   E = ☐   F = ☐

2 Mark the numbers on the number line.

A = 500   B = 6000   C = 8100   D = 1600   E = 3500   F = 9900

3 In Question 2, all the numbers A, B, C, D, E and F are between 0 and 10 000. Compare the distances of each number from these two numbers on the number line, and write the correct letters in each space.

(a) The numbers that are nearer to 0 are _____.

(b) The numbers that are nearer to 10 000 are _____.

4 True or false? (Put a ✓ for true and a ✗ for false in each box.)
You may use number lines to help find the answers.

(a) Rounding 19 to the nearest 10, the result is 20. . . . . ☐

(b) Rounding 19 to the nearest 100, the result is 200. . . . . ☐

(c) Rounding 19 to the nearest 1000, the result is 2000. . . . . ☐

(d) Rounding 5890 to the nearest 1000, the result is 6000. . . . . . . ☐

(e) Rounding 8850 to the nearest 1000, the result is 9000. . . . . . ☐

(f) Rounding 9549 to the nearest 1000, the result is 10 000. . . . ☐

5 Round the numbers to their nearest numbers as indicated.

(a) Round to the nearest 10.

(b) Round to the nearest 100.

(c) Round to the nearest 1000.

Number	To the nearest 10	Number	To the nearest 100	Number	To the nearest 1000
35		51		20	
9		325		199	
91		956		501	
501		1501		2300	
199		3020		4708	
2021		5050		7499	
4093		8116		8499	
9999		9999		9999	

6 I am a number. When I am rounded to the nearest 10, the result is 50. When I am rounded to the nearest 100, the result is 0. Write down all the numbers that I can possibly be.

Answer: _____

7 I am a number. When I am rounded to the nearest 10, the result is 7000. When I am rounded to the nearest 100, the result is 7000. When I am rounded to the nearest 1000, the result is still 7000. I am not 7000. Write down all the numbers that I can possibly be.

Answer: _____

2.4 Addition with 4-digit numbers (1)

Learning objective Add numbers with up to 4 digits

Basic questions

1 Calculate with reasoning.

(a) Kaya's method.

(i) 3130 + 4216 = ☐

Thousands + Thousands: 3000 + 4000 = ☐

Hundreds + Hundreds: 100 + 200 = ☐

Tens + Tens: 30 + 10 = ☐

Ones + Ones: 0 + 6 = ☐

☐ + ☐ + ☐ + ☐ = ☐

(ii) 2068 + 1107 = ☐

Thousands + Thousands: ☐ + ☐ = ☐

Hundreds + Hundreds: ☐ + ☐ = ☐

Tens + Tens: ☐ + ☐ = ☐

Ones + Ones: ☐ + ☐ = ☐

☐ + ☐ + ☐ + ☐ = ☐

(b) Amir's method.

(i) 7132 + 1454 = ▢

Ones + Ones: ▢ + ▢ = ▢

Tens + Tens: ▢ + ▢ = ▢

Hundreds + Hundreds: ▢ + ▢ = ▢

Thousands + Thousands: ▢ + ▢ = ▢

▢ + ▢ + ▢ + ▢ = ▢

(ii) 3063 + 4705 = ▢

Ones + Ones: ▢ + ▢ = ▢

Tens + Tens: ▢ + ▢ = ▢

Hundreds + Hundreds: ▢ + ▢ = ▢

Thousands + Thousands: ▢ + ▢ = ▢

▢ + ▢ + ▢ + ▢ = ▢

(c) (i) Joe's method.

3024 + 2509

= 3024 + 2000 + 500 + 9

= ▢ + ▢ + ▢

= ▢ + ▢

= ▢

(ii) 3175 + 1423

= 3175 + ☐ + ☐ + ☐ + ☐

= ☐ + ☐ + ☐ + ☐

= ☐ + ☐ + ☐

= ☐ + ☐

= ☐

(d) Erin's method.

(i) 4130 + 3512

= 4130 + 2 + 10 + 500 + 3000

= ☐ + ☐ + ☐ + ☐

= ☐ + ☐ + ☐

= ☐ + ☐

= ☐

(ii) 1782 + 7219

= 1782 + ☐ + ☐ + ☐ + ☐

= ☐ + ☐ + ☐ + ☐

= ☐ + ☐ + ☐

= ☐ + ☐

= ☐

2 Use your preferred method to calculate. Show your working.

(a) 5136 + 8121 = ⬚

(b) 4228 + 2436 = ⬚

(c) 6750 + 3128 = ⬚

3 Use the six number cards below to form three addition sentences each adding two 4-digit numbers. Then use your preferred method to work out the answers.

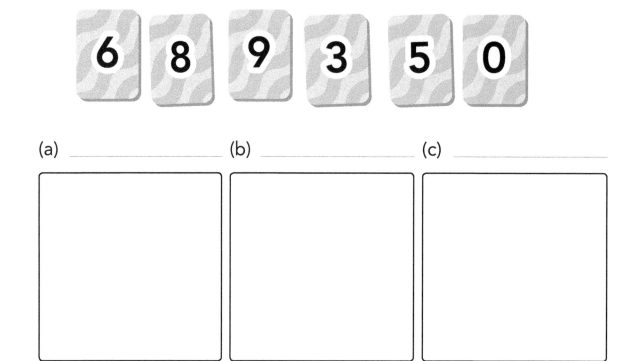

(a) _____   (b) _____   (c) _____

4 Write the number sentences and calculate.

(a) What number is 1000 more than 1554?

Answer: _____

(b) What number is 1000 less than 5528?

Answer: _____

(c) One addend is 2139 and the other addend is 3324.
What is the sum?

Answer: _____

Challenge and extension question

5 Fill in the boxes.

(a) The sum of the smallest 4-digit number and the smallest 3-digit number is a ▢ -digit number.

(b) The sum of the greatest 4-digit number and the smallest 3-digit number is a ▢ -digit number.

(c) The sum of two 4-digit numbers can be a ▢ -digit number or a ▢ -digit number.

2.5 Addition with 4-digit numbers (2)

Learning objective Add numbers with up to 4 digits

Basic questions

1 Use the column method to calculate.

(a) 3500 + 1234 = [   ]

```
   3 5 0 0
 + 1 2 3 4
 _____
```

(b) 308 + 7123 = [   ]

```
     3 0 8
 + 7 1 2 3
 _____
```

(c) 7095 + 225 = [   ]

```
   7 0 9 5
 +   2 2 5
 _____
```

(d) 5907 + 180 = [   ]

(e) 5372 + 1043 + 2301 = ☐

2 Complete the table.

Addend	1327	3204	584	1178	1257	9178
Addend	1150	2328	4265	7433	4465	822
Sum						

3 Is each calculation correct? Underneath, put a ✓ if it is and a ✗ if it is not and make the corrections.

(a)
```
    3 8 0 1
  +   2 0 3
  ─────────
    5 8 3 1
```
☐

(b)
```
    7 0 9 1
  + 2 0 6 1
  ─────────
    9 0 5 2
```
☐

(c)
```
      5 2 4
  +   9 8 2
  ─────────
  1 4 1 0 6
```
☐

Corrections:

4 Now try these.

(a) 32 145 + 25 134 =

```
    3   2   1   4   5
+   2   5   1   3   4
_____

_____
```

(b) 7408 + 6925 =

```
    7   4   0   8
+   6   9   2   5
_____

_____
```

5 Application problems.

(a) Adam walked 1200 metres in the morning and 1300 metres in the afternoon. How far did he walk that day?

Answer: _____

(b) Shireen earned £2580 in July and £2880 in August. How much did she earn in the two months?

Answer: _____

(c) A factory consumed 3326 kilowatts of electricity in the winter season in a year and 2539 kilowatts in the spring season. What is the total usage of the electricity in these two seasons?

Answer: _____

Challenge and extension question

6 Write suitable numbers in the boxes to make the calculations correct.

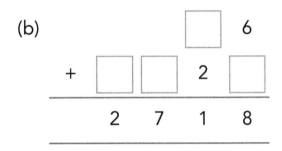

(a)

		□	9	□	□
+		3	□	3	3
		7	7	5	5

(b)

			□	6
		□	2	□
+	□			
	2	7	1	8

(c)

		5	3	7	□
+	□	□	□	□	2
	3	3	3	0	0

2.6 Subtraction with 4-digit numbers (1)

 Learning objective Subtract numbers with up to 4 digits

 Basic questions

1 Calculate with reasoning.

(a) Dillon's method.

    (i) 9746 − 5443 = ☐

        Subtract thousands first: 9746 − 5000 = ☐

        Then subtract hundreds: ☐ − ☐ = ☐

        Then subtract tens: ☐ − ☐ = ☐

        Finally subtract ones: ☐ − ☐ = ☐

    (ii) 6957 − 3356 = ☐

        Subtract thousands first: ☐ − ☐ = ☐

        Then subtract hundreds: ☐ − ☐ = ☐

        Then subtract tens: ☐ − ☐ = ☐

        Finally subtract ones: ☐ − ☐ = ☐

(b) Amadi's method.

(i) 5487 − 3245 = [____]

Thousands − Thousands: | 5000 | − | 3000 | = [____]

Hundreds − Hundreds: | 400 | − | 200 | = [____]

Tens − Tens: | 80 | − | 40 | = [____]

Ones − Ones: | 7 | − | 5 | = [____]

[____] + [____] + [____] + [____] = [____]

(ii) 7965 − 5342 = [____]

Thousands − Thousands: [____] − [____] = [____]

Hundreds − Hundreds: [____] − [____] = [____]

Tens − Tens: [____] − [____] = [____]

Ones − Ones: [____] − [____] = [____]

[____] + [____] + [____] + [____] = [____]

(c) Joe's method.

(i) 5306 − 2285

= | 5306 | − | 2000 | − | 200 | − | 80 | − | 5 |

= [____] − [____] − [____] − [____]

= [____] − [____] − [____]

= [____] − [____]

= [____]

(ii) 6238 – 1729

= [6238] – [ ] – [ ] – [ ] – [ ]

= [ ] – [ ] – [ ] – [ ]

= [ ] – [ ] – [ ]

= [ ] – [ ]

= [ ]

(d) Ava's method.

(i) 9008 – 3627

= [9008] – [7] – [20] – [600] – [3000]

= [ ] – [ ] – [ ] – [ ]

= [ ] – [ ] – [ ]

= [ ] – [ ]

= [ ]

(ii) 8536 – 3097

= [8536] – [ ] – [ ] – [ ]

= [ ] – [ ] – [ ]

= [ ] – [ ]

= [ ] – [ ]

2 Use your preferred method to calculate. Show your working.

(a) 7759 – 4325 = ⬚

(b) 5000 – 2169 = ⬚

(c) 8439 – 7346 = ⬚

(d) 9557 – 6262 = ⬚

(e) 4821 – 2656 = ⬚

(f) 6668 – 1095 = ⬚

3 Use these six numbers to form subtraction sentences that each subtract a 4-digit number from another 4-digit number. Then use your preferred method to calculate.

(a) _____ (b) _____ (c) _____

4 Write the number sentences and calculate.

(a) A number is 3415 less than 5032. What is the number?

Answer: _____

(b) The minuend is 9418 and the subtrahend is 2280. What is the difference?

Answer: _____

5 The mass of one box of apples and one box of pears is 7450 grams. The mass of two boxes of apples and one box of pears is 9550 grams. Write the number sentences and calculate.

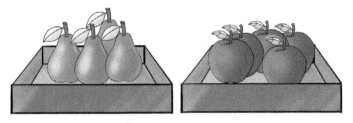

(a) What is the mass of one box of apples?

Answer: _____

(b) What is the mass of one box of pears?

Answer: _____

2.7 Subtraction with 4-digit numbers (2)

Learning objective Subtract numbers with up to 4 digits

Basic questions

1 Use the column method to calculate.

(a) 5685 − 2453 = [____]

```
    5 6 8 5
  − 2 4 5 3
  _____
```

(b) 3149 − 778 = [____]

```
    3 1 4 9
  −   7 7 8
  _____
```

(c) 5151 − 4279 = [____]

```
    5 1 5 1
  − 4 2 7 9
  _____
```

(d) 7435 − 3267 = [____]

(e) 5507 − 3368 = ⬚   (f) 1001 − 703 + 5689 = ⬚

2 Is each calculation correct? Put a ✓ if it is and a ✗ if it is not and make the corrections.

(a)
```
    3  4  5  5
 -  1  1  0  9
 ─────────────
    2  3  5  6
```
⬚

(b)
```
    6  5  1  7
 -  3  2  2  6
 ─────────────
    3  3  3  1
```
⬚

(c)
```
    8  0  0  5
 -  7  3  9  6
 ─────────────
    1  6  1  9
```
⬚

Corrections:

3 Fill in the table.

Minuend	8738	9564	4657	6771	5316	7008
Subtrahend	1534	5268	2476	6635	2138	6569
Difference						

4 Write suitable numbers in the boxes to make the calculations correct.

(a)
```
    □  5  □  □
  -    4  □  5  9
  _____
    3  5  3  8
```

(b)
```
    □  4  0  □
  -    7  □  □  5
  _____
    1  2  0  5
```

(c)
```
    6  7  □  3
  -      □  3  □
  _____
    6  3  4  8
```

5 Give this a try.

(a) 65 077 − 43 015 = ☐

```
    6  5  0  7  7
  - 4  3  0  1  5
  _____
```

(b) 11 078 − 8569 = ☐

```
    1  1  0  7  8
  -    8  5  6  9
  _____
```

6 The flight distances from London to Amsterdam, Beijing and New York are 358 km, 8161 km and 5586 km, respectively.

(a) What is the difference between the distance from London to Amsterdam and that from London to Beijing?

Answer: _____

(b) On a business trip, Paul flew from New York to London on the first day and then from London to Beijing on the second day. Which day did he fly further and by how much? What was the total flight distance he flew in the two days?

Answer: _____

Challenge and extension question

7 What number does ■ stand for in the subtraction below?

$$8\blacksquare 62$$
$$-7955$$

(a) If the difference is a 4-digit number, the ■ must stand for ☐.

(b) If the difference is a 3-digit number, the greatest possible number the ■ can stand for is ☐.

2.8 Estimating and checking answers using inverse operations

Learning objective Estimate and check answers to a calculation

Basic questions

1 Work out the answers mentally. What do you find?

(a) 5200 + 800 =

(b) 3100 − 900 =

(c) 6700 + 2000 =

(d) 9020 − 220 =

(e) 6000 − 800 =

(f) 3100 − 2200 =

(g) 8700 − 2000 =

(h) 9020 − 8800 =

(i) 6000 − 5200 =

(j) 2200 + 900 =

(k) 8700 − 6700 =

(l) 8800 + 220 =

2 Round the following numbers to the nearest 10, 100 and 1000.

	2132	5522	4590	6705	1848	8999
Nearest 10						
Nearest 100						
Nearest 1000						

3 Estimate to the nearest 100 and then calculate.

(a) 2012 + 1689

Estimate:

Calculate:

(b) 5431 + 3309

Estimate:

Calculate:

(c) 5996 + 2992 + 889

Estimate:

Calculate:

(d) 5674 − 2318

Estimate:

Calculate:

(e) 7883 − 5479 − 2078

Estimate:

Calculate:

(f) 9989 − 2994 + 3030

Estimate:

Calculate:

4 Estimate to the nearest 1000 first and then calculate.

(a) 1308 + 4117

Estimate: ☐

Calculate:

(b) 3657 + 6329

Estimate: ☐

Calculate:

(c) 5291 + 1428 + 3049

Estimate: ☐

Calculate:

(d) 9417 − 7206

Estimate: ☐

Calculate:

(e) 6529 − 780 − 2115

Estimate: ☐

Calculate:

(f) 7658 − 2300 + 3100

Estimate: ☐

Calculate:

5 Use the column method to calculate and then use inverse operations to check your answers. If an answer is wrong, correct it. Remember, addition and subtraction are inverse operations of each other. The first one has been done for you.

(a)
```
    3  4  0  5
+   1  5  3  5
_____
    4  9  4  0
```

Check: Does 4940 – 1535 = 3405?

Yes, it works.

(b)
```
    5  5  4  8
+   4  3  7  1
_____
    9  8  1  9
```

Check:

(c)
```
    9  2  0  8
-   3  2  5  7
_____
    5  0  5  1
```

Check:

(d)
```
    8  3  9  9
+      6  9  9
_____
    8  9  8  8
```

Check:

(e)
```
    4  0  3  2
-   2  3  9  1
_____
    1  6  4  1
```

Check:

(f)
```
 1  0  0  0  0
-      4  0  7  5
_____
    5  0  2  5
```

Check:

6 Application problems.

(a) An aquarium has two water tanks. The larger one has a capacity of 5230 litres and the smaller one has a capacity of 1755 litres. What is the difference between the capacities of the water tanks? What is their total capacity?

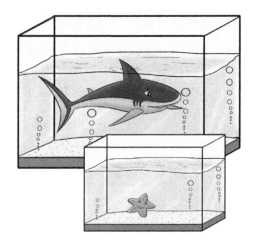

Answer: _____

(b) A school plans to spend £5850 on computers and £1750 on printers.

(i) The budget for the items is £7000. Estimate whether this is enough for purchasing the computers and printers.

Answer: _____

(ii) If the budget is not sufficient, how much more money is needed to purchase the computers and printers?

Answer: _____

7 A number consists of four digits: 8, 8, 6 and 2.

(a) If you add the number and 2500, the result is between 5000 and 6000. This number is ⬚ .

(b) If you subtract 2500 from the number, the result is between 3500 and 4000. This number is ⬚ .

Chapter 2 test

1 Calculate mentally and then write the answers.

(a) 3000 + 2000 = ⬚

(b) 8690 − 690 = ⬚

(c) 5240 + 4750 = ⬚

(d) 2500 + 4500 = ⬚

(e) 7190 − 4190 = ⬚

(f) 6070 − 3010 = ⬚

(g) 9540 − 9450 = ⬚

(h) 7000 + ⬚ = 10 000

(i) ⬚ − 3220 = 5000

2 Write the number that each letter stands for.

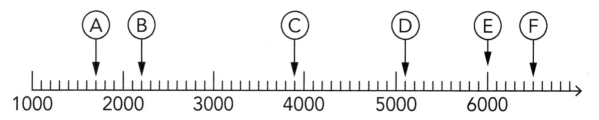

A = ⬚

B = ⬚

C = ⬚

D = ⬚

E = ⬚

F = ⬚

3 Count and complete number patterns.

(a) 505, 510, 515, ⬚ , ⬚ , 530

(b) 505, 605, 705, ⬚ , ⬚ , 1005

(c) 5000, 6000, 7000, ⬚ , ⬚ , 10 000

(d) 200, 175, 150, ⬚ , ⬚

(e) 12, 24, 36, ⬚ , ⬚ , ⬚ , 84

4 Complete the place value charts.

(a) 3019

Thousands	Hundreds	Tens	Ones

(b) 7909

Thousands	Hundreds	Tens	Ones

(c) 1535

Thousands	Hundreds	Tens	Ones

(d) 1000

Ten Thousands	Thousands	Hundreds	Tens	Ones

5 Use the column method to calculate. Check the answers to the questions marked with *.

(a) *5939 + 811 = ☐

(b) *10 000 − 8917 = ☐

(c) *3892 + 4308 = ☐

(d) 8917 − 7289 + 1023 = ☐

(e) 5003 + 2085 − 3999 = ☐

6 Estimate to the nearest 1000 first. Then calculate.

(a) 3771 + 2117

Estimate: ☐

Calculate:

(b) 1181 + 8807 − 3500

Estimate: ☐

Calculate:

(c) 8512 − 3296

Estimate: [   ]

Calculate:

(d) 9389 − 7209 + 2311

Estimate: [   ]

Calculate:

7 Write the number sentences and then find the answers.

(a) What number is 1000 less than 10 000?

Number sentence: _____

Answer: _____

(b) The sum of two addends is 8569. One addend is 3378. What is the other addend?

Number sentence: _____

Answer: _____

(c) The subtrahend is 8288 and the difference is 3009. What is the minuend?

Number sentence: _____

Answer: _____

8 A fruit and vegetable shop earned £8050 in July and £9558 in August by selling local and seasonal foods.

(a) How much did the shop earn in these two months?

Number sentence: _____

Answer: _____

(b) How much more did it earn in August than in July?

Number sentence: _____

Answer: _____

9 There are 1337 pupils in Year 2, 1021 pupils in Year 3 and 1368 pupils in Year 4. How many pupils are there in total in these three year groups?

Number sentence: _____

Answer: _____

10 The flight distance from London to
Rome is 1434 km and that from
London to Oslo is 1157 km.
Kara plans to fly from Rome to London
on the first day and then from
London to Oslo on the second day.

(a) On which day will Kara fly further, and by how much?

Answer: _____

(b) What is the total flight distance Kara will travel in these
two days?

Answer: _____

Chapter 3 Multiplying by a 2-digit number

3.1 Multiplying whole tens by a 2-digit number

 Learning objective Multiply numbers by multiples of 10

 Basic questions

1 Calculate with reasoning.

(a)
4		

40	× 12 =	

400		

(b)
450		

45	× 7 =	

4500		

Fill in the boxes.

(c) $40 \times 12 = 10 \times 4 \times 12 = 10 \times \boxed{} = \boxed{}$

(d) $450 \times 7 = 10 \times 45 \times 7 = 10 \times \boxed{} = \boxed{}$

(e) $400 \times 12 = 100 \times 4 \times 12 = 100 \times \boxed{} = \boxed{}$

(f) $4500 \times 7 = 100 \times 45 \times 7 = 100 \times \boxed{} = \boxed{}$

2 Look carefully and then write the answers.

(a) $11 \times 2 = \boxed{}$

$110 \times 2 = \boxed{}$

$11 \times 200 = \boxed{}$

$110 \times 20 = \boxed{}$

(b) $3 \times 15 = \boxed{}$

$30 \times 15 = \boxed{}$

$3 \times 1500 = \boxed{}$

$300 \times 15 = \boxed{}$

(c) $45 \times 6 = \boxed{}$

$450 \times 6 = \boxed{}$

$450 \times 60 = \boxed{}$

$4500 \times 6 = \boxed{}$

3 Complete the following mental calculations.

(a) $63 \times 60 = \boxed{}$

(b) $100 \times 30 = \boxed{}$

(c) $4 \times 250 = \boxed{}$

(d) $17 \times 20 = \boxed{}$

(e) $42 \times 30 = \boxed{}$

(f) $50 \times 18 = \boxed{}$

(g) $75 \times 30 = \boxed{}$

(h) $160 \times 9 = \boxed{}$

(i) $6 \times 1500 = \boxed{}$

(j) $2 \times 50 = \boxed{}$

(k) $88 \times 30 = \boxed{}$

(l) $700 \times 17 = \boxed{}$

(m) $13 \times 500 = \boxed{}$

(n) $8 \times 400 = \boxed{}$

(o) $6 \times 110 = \boxed{}$

(p) $5400 \times 6 = \boxed{}$

4 Write the number sentences and then calculate.

(a) What is 12 times 60?

_____

(b) What is the sum of 800 twenty-fives?

_____

5 Write >, < or = in each ◯ without calculation.

(a) 250 × 40 ◯ 40 × 250

(b) 1300 × 4 ◯ 13 × 40

(c) 6500 × 330 ◯ 650 × 330

(d) 47 × 210 ◯ 470 × 210

6 Think carefully and calculate the following.

> When you work on a number sentence with brackets,
> perform the calculations inside the brackets first.

(a) 34 × 6 = ☐

30 × 6 = ☐

4 × 6 = ☐

(30 × 6) + (4 × 6) = ☐

(b) 52 × 7 = ☐

50 × 7 = ☐

2 × 7 = ☐

(50 × 7) + (2 × 7) = ☐

7 True or false? (Put a ✓ for true and a ✗ for false in each box.)

(a) 50 × 21 = 50 × (20 + 1) = 50 × 20 + 1 = 1000 + 1 = 1001 . . . . ☐

(b) 50 × 21 = 50 × (20 + 1) = 50 × 20 + 50 × 1 = 1000 + 50 = 1050 . . . ☐

(c) 29 × 30 = (30 − 1) × 30 = 30 × 30 − 1 × 30 = 900 − 30 = 870 . . . . . ☐

(d) Given three numbers A, B and C, we have:

A × (B + C) = A × B + A × C . . . . ☐

A × (B − C) = A × B − A × C . . . . ☐

8 Calculate smartly. The first one has been done for you.

(a) $35 \times 7 + 35 \times 3$

$= 35 \times (7 + 3)$

$= 35 \times 10$

$= 350$

(b) $159 \times 32 - 59 \times 32$

$= (159 - 59) \times 32$

$=$

$=$

(c) $656 \times 62 - 56 \times 62$

(d) $62 \times 98 + 38 \times 98$

9 Two boxes of soft drinks each contain 24 bottles. What is the total cost of all the bottles? Use two different methods to find the answer.

50p/bottle

70p/bottle

Method 1:

Method 2:

Challenge and extension question

10 In each number sentence below, fill in the two boxes with the same 2-digit number so the equation is true. The first one has been done for you.

(a) 5 × ⬚50⬚ = 2⬚50⬚   (b) 6 × ⬚ ⬚ = 3⬚ ⬚   (c) 9 × ⬚ ⬚ = 4⬚ ⬚

3.2 Multiplying a 2-digit number by a 2-digit number (1)

Learning objective Use different methods to multiply 2-digit numbers together

Basic questions

1 Calculate mentally and then write the answers.

> When you work on a number sentence like this, perform all the multiplications/divisions first and then perform the additions/subtractions.

(a) $16 \times 10 = \boxed{}$

(b) $16 \times 5 = \boxed{}$

(c) $16 \times 10 + 16 \times 5 = \boxed{}$

(d) $16 \times 10 + 16 \times 5 = 16 \times \boxed{}$

(e) $24 \times 10 = \boxed{}$

(f) $24 \times 5 = \boxed{}$

(g) $24 \times 10 + 24 \times 5 = \boxed{}$

(h) $24 \times 10 + 24 \times 5 = 24 \times \boxed{}$

(i) $33 \times 10 = \boxed{}$

(j) $33 \times 5 = \boxed{}$

(k) $33 \times 10 + 33 \times 5 = \boxed{}$

(l) $33 \times 10 + 33 \times 5 = 33 \times \boxed{}$

2 Estimate first and then calculate.

(a) What is 13 twenty-fours?

Estimate: The answer is between ☐ and ☐.

Calculate: 13 × 20 = ☐

13 × 4 = ☐

13 × 24 = ☐

(b) What is 31 sixty-twos?

Estimate: The answer is between ☐ and ☐.

Calculate: 31 × 60 = ☐

31 × 2 = ☐

31 × 62 = ☐

3 Calculate with different methods.

(a)
48 × 25

= 12 × 4 × ☐

=

=

(b)
48 × 25

= 40 × ☐ + 8 × ☐

=

=

(c)
48 × 25

= 50 × ☐ − 2 × ☐

=

=

Do you know any other methods to calculate?

4 Calculate.

(a)
$$19 \times 21$$
$$= 19 \times 20 + 19 \times \boxed{}$$
$$=$$
$$=$$

(b)
$$33 \times 77$$
$$= 33 \times \boxed{} + 33 \times \boxed{}$$
$$=$$
$$=$$

(c)
$$51 \times 63$$
$$=$$
$$=$$
$$=$$

5 Write the number sentences and then calculate.

(a) What is the product of 11 and 55?

_____

(b) How much more is 550 than 19 times 19?

_____

 Challenge and extension question

6 In the calculation below, each letter represents one number. Different letters represent different numbers. What numbers do they represent in order to make the calculation correct?

$$
\begin{array}{r}
1\ A\ B\ C\ D\ E \\
\times \qquad\qquad 3 \\
\hline
A\ B\ C\ D\ E\ 1 \\
\end{array}
$$

A = ☐   B = ☐   C = ☐   D = ☐   E = ☐

3.3 Multiplying a 2-digit number by a 2-digit number (2)

Learning objective Use formal written methods to multiply 2-digit numbers together

Basic questions

1 Calculate mentally and then write the answers.

(a) 180 + 135 = ☐

(b) 53 × 7 = ☐

(c) 640 − 480 = ☐

d) 15 × 2 × 5 = ☐

(e) 580 + 472 = ☐

(f) 70 ÷ 70 + 2 = ☐

(g) 60 × 700 = ☐

(h) 95 ÷ 5 = ☐

(i) 220 × 4 = ☐

(j) 25 × 80 = ☐

(k) 12 × 55 = ☐

(l) 3000 ÷ 4 = ☐

2 Fill in the boxes to complete the column calculation.

48 × 65 = ☐

```
        4   8
  ×     6   5
  _____
    ☐   ☐   ☐    → First multiply 48 and ☐ in the ones place.
  ☐   ☐   ☐  0   → Then multiply 48 and ☐ in the tens place.
  _____
  ☐   ☐   ☐   ☐
```

3 Try it on your own. Use the column method to calculate the following.

(a) 13 × 22 =

(b) 75 × 99 =

(c) 63 × 48 =

4 Where are the mistakes? Identify them first and then correct them.

(a) 44 × 55 = 440

(b) 37 × 12 = 111

(c) 26 × 98 = 442

		4	4
×		5	5
	2	2	0
	2	2	0
	4	4	0

My correction:

		3	7
×		1	2
		7	4
		3	7
	1	1	1

My correction:

		2	6
×		9	8
	2	0	8
	2	3	4
	4	4	2

My correction:

5 Write the number sentences and then calculate.

(a) What is the product of two eighty-nines?

(b) What is the product of the two greatest 2-digit numbers?

6 Application problems.

(a) In a charity fundraiser, Year 4 pupils donated £99 and the amount their teachers donated was 33 times as much. How much did the pupils and teachers donate altogether?

(b) In a tree-planting activity organised by a school, 53 participating pupils were from Year 3. The number of pupils from Year 4 was the same. The number of pupils who took part from Year 5 was twice the total number of the pupils from Year 3 and Year 4. How many pupils in Year 5 took part in the tree-planting activity?

Challenge and extension question

7 Calculate and try to remember the answers. Find the patterns.

(a) $11 \times 11 =$

(b) $11 \times 11 =$

(c) $12 \times 12 =$

(d) $12 \times 11 =$

(e) $13 \times 13 =$

(f) $13 \times 11 =$

(g) $14 \times 14 =$

(h) $14 \times 11 =$

(i) $15 \times 15 =$

(j) $15 \times 11 =$

(k) $16 \times 16 =$

(l) $16 \times 11 =$

(m) $17 \times 17 =$

(n) $17 \times 11 =$

(o) $18 \times 18 =$

(p) $18 \times 11 =$

(q) $19 \times 19 =$

(r) $19 \times 11 =$

3.4 Multiplying a 3-digit number by a 2-digit number (1)

 Learning objective Use different methods to multiply 3-digit numbers by 2-digit numbers

 Basic questions

1 Calculate mentally and then write the answers.

(a) $125 \times 3 =$ ☐

(b) $125 \times 5 =$ ☐

(c) $125 \times 7 =$ ☐

(d) $125 + 375 =$ ☐

(e) $125 + 250 =$ ☐

(f) $125 + 500 =$ ☐

(g) $125 + 625 =$ ☐

(h) $125 \times 9 =$ ☐

(i) $125 \times 4 =$ ☐

(j) $125 \times 6 =$ ☐

(k) $125 \times 8 =$ ☐

(l) $125 \times 11 =$ ☐

2 Estimate first and then calculate.

(a) What is 112 forty-sixes?

Estimate: The answer is between ☐ and ☐ .

Calculate: $112 \times 40 =$ ☐

$112 \times 6 =$ ☐

$112 \times 46 =$ ☐

(b) What is 229 twenty-ones?

Estimate: The answer is between ☐ and ☐.

Calculate: 229 × 20 = ☐

229 × 1 = ☐

229 × 21 = ☐

3 Fill in the boxes to complete the column calculation.

418 × 65 = ☐

```
      4   1   8
  ×           6   5
  _____
  ☐   ☐   ☐   ☐        → First multiply 418 by 5 in the ones place.
  ☐   ☐   ☐   ☐   0    → Then multiply 418 by 6 in the tens place.
  ☐   ☐   ☐   ☐   ☐
  _____
```

4 Fill in the spaces to show how the column calculations have been done.

(a)

```
      1   2   3
  ×       4   5
  _____
      6   1   5    ...  123  ×  5 ones
  4   9   2   0    ...  123  ×  ____
  _____
  5   5   3   5
```

(b)

```
        3   5   7
  ×         8   9
  _____
        3   2   1   3   ...  [   ]  ×  _____
    2   8   5   6   0   ...  [   ]  ×  _____
  _____
    3   1   7   7   3
  _____
```

5 Use the column method to calculate.

(a) 327 × 11 =   (b) 256 × 32 =   (c) 555 × 99 =

6 Write the number sentences and then calculate.

(a) What is the sum of 222 fifty-fives?

_____

(b) What is the product of the greatest 2-digit number multiplied by the greatest 3-digit number?

_____

7 Application problem.

£329	£300 less than 12 times the price of the microwave

How much do the two items cost in total?

Challenge and extension question

8 (a) It took Alvin 100 seconds to walk from the first floor to the fifth floor. How many seconds will it take him to walk to the eleventh floor if he continues at the same pace?

(b) A convoy of 52 trucks stopped for inspection. Each truck was 4 metres long. The distance between two trucks was 6 metres.

The total length of the convoy was ☐ metres.

3.5 Multiplying a 3-digit number by a 2-digit number (2)

 Learning objective Use formal written methods to multiply 3-digit numbers by 2-digit numbers

 Basic questions

1 Calculate mentally and then write the answers.

(a) $14 \times 2 =$ ☐

(b) $140 \times 2 =$ ☐

(c) $140 \times 20 =$ ☐

(d) $6 \times 5 =$ ☐

(e) $60 \times 5 =$ ☐

(f) $60 \times 500 =$ ☐

(g) $13 \times 3 =$ ☐

(h) $13 \times 30 =$ ☐

(i) $130 \times 300 =$ ☐

2 Write a suitable number in each box.
The first one has been done for you.

(a) 28×500

$= 28 \times 5 \times \boxed{100}$

$= \boxed{140} \times \boxed{100}$

$= \boxed{14\,000}$

(b) 220×400

$= 220 \times 4 \times \boxed{}$

$= \boxed{} \times \boxed{}$

$= \boxed{}$

(c) 560×120

$= 560 \times 12 \times \boxed{}$

$= \boxed{} \times \boxed{}$

$= \boxed{}$

3 Fill in the box based on 32 × 45 = 1440.
The first one has been done for you.

(a) 320 × $\boxed{4500}$ = 1 440 000

(b) 32 × $\boxed{}$ = 1 440 000

(c) $\boxed{}$ × 450 = 1 440 000

(d) $\boxed{}$ × 45 = 1 440 000

4 Multiple choice questions. (For each question, choose the correct answer and write the letter in the box.).

> Remember, when two numbers are multiplied, the result is called the product, and the two numbers are called the factors.
>
> product = factor × factor

(a) In calculating 246 × 75, the product of multiplying 246 by the digit 7 in the tens place of the factor 75 is: $\boxed{}$

A. 17 220   **B.** 1722   **C.** 1 722 000   **D.** 172 200

(b) When 225 is added 15 times, the sum is: $\boxed{}$
(Clue: count 225 + 225 as 225 being added twice.)

A. 33 750   **B.** 337 500   **C.** 3 375 000   **D.** 3375

(c) When one factor has 2 zeros at the end and the other factor has 1 zero at the end, then the product has $\boxed{}$ zeros at the end.

A. 2   **B.** 3   **C.** at least 3   **D.** 4

(d) The product of two factors is 120. If one factor is doubled and the other is increased to 5 times its original value, then the new product is: $\boxed{}$

A. 1200   **B.** 360   **C.** 900   **D.** 180

5 Use the column method to calculate.

(a) 349 × 57 =

(b) 608 × 22 =

(c) 4200 × 150 =

(d) 39 × 208 =

(e) 26 × 737 =

(f) 404 × 88 =

6 Write the number sentences and then calculate.

(a) A is 200, B is 12 times as large as A and C is 12 times as large as B. What is C?

(b) The divisor is 160. Both the quotient and the remainder are 50. What is the dividend?

Challenge and extension question

7 There are 20 classes in a school and each class has 32 pupils. A cinema has 1000 seats. Is this enough to seat all the pupils? If so, how many seats will be left available?

3.6 Dividing 2-digit or 3-digit numbers by tens

Learning objective Use different methods to divide 3-digit numbers by multiples of ten

Basic questions

1 Calculate mentally and then write the answers.

(a) 80 ÷ 2 = ☐   (e) 150 ÷ 5 = ☐

(b) 80 ÷ 20 = ☐   (f) 150 ÷ 50 = ☐

(c) 840 ÷ 7 = ☐   (g) 630 ÷ 9 = ☐

(d) 840 ÷ 70 = ☐   (h) 630 ÷ 90 = ☐

2 Write the greatest number you can in each box to make the statement true.

(a) 50 × ☐ < 157   (b) ☐ × 30 < 231   (c) 70 × ☐ < 369

(d) ☐ × 40 < 369   (e) 80 × ☐ < 508   (f) ☐ × 90 < 396

3 Think carefully and then write the missing numbers in the boxes. The first one has been done for you.

(a) 92 ÷ 40 = ?

Think: In 9 ÷ 4, the quotient is ☐2☐.

In 92 ÷ 40, the quotient is ☐2☐.

92 − 40 × ☐2☐ = ☐12☐

92 ÷ 40 = ☐2☐ r ☐12☐

(b) 362 ÷ 40 = ?

Think: In 36 ÷ 4, the quotient is ☐.

In 362 ÷ 40, the quotient is ☐.

362 − 40 × ☐ = ☐

362 ÷ 40 = ☐ r ☐

(c) 75 ÷ 20 = ?

Think: How many 20s are there in 75?

20 × ☐ < 75

20 × ☐ > 75

There are ☐ 20s in 75.

Then: in 75 ÷ 20, the quotient is ☐.

So: 75 ÷ 20 = ☐ r ☐

(d) 256 ÷ 30 = ?

Think: How many 30s are there in 256?

30 × ☐ < 256

30 × ☐ > 256

There are ☐ 30s in 256.

Then: in 256 ÷ 30, the quotient is ☐.

So: 256 ÷ 30 = ☐ r ☐

4 Use the column method to calculate. The first one has been done for you.

(a) $96 \div 60 = 1 \text{ r } 36$

(b) $145 \div 20 =$

(c) $572 \div 80 =$

(d) $360 \div 70 =$

(e) $455 \div 50 =$

(f) $666 \div 90 =$

5 Write the number sentences and then calculate.

(a) 292 is divided by 60. What is the quotient? What is the remainder?

(b) What are the quotient and remainder of the greatest 2-digit number divided by 20?

6   There are 720 pupils in a school. If they stand in different numbers of rows, as indicated in the table below, how many pupils are there in each row? Write your answers in the table.

Number of rows	10	20	30	40	60	80
Number of pupils in each row						

Challenge and extension question

7   (a) Subtract a whole tens number from 480 and then divide the difference by the whole tens number. The quotient is 5.

This whole tens number is _____.

(b) In a long distance run, Aliya was 70 metres ahead of Joe. Shea was 40 metres behind Lily. Joe was 30 metres ahead of

Shea. The first runner was _____ and the third runner

was _____.

(c) When a dividend is divided by a divisor, the quotient is 7 and the remainder is 3. If the sum of the divisor, the dividend, the quotient and the remainder is 85, then the dividend

is _____.

(d) Two houses are 250 metres apart. Mr Wood plants 49 trees in a straight line between the two houses. The distance from one tree to the next is equal.

There are _____ metres between each tree and the next.

3.7 Practice and exercise

Learning objective Use different methods to multiply and divide 3-digit numbers and 2-digit numbers

Basic questions

1 Calculate mentally and then write the answers.

(a) $16 \times 30 =$

(b) $34 \times 5 \div 10 =$

(c) $11 \times 11 \times 0 =$

(d) $4800 \div 20 =$

(e) $72 \times 5 - 20 =$

(f) $100 \div 20 + 100 =$

(g) $9 \times$ _____ $= 9090$

(h) $1000 \div$ _____ $= 125$

2 Use the column method to calculate. Check the answer to the question marked with*.

(a) $567 \times 11 =$

(b) $313 \times 32 =$

(c) $2030 \times 420 =$

(d) * $2551 \div 30 =$

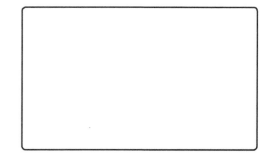

3 Work these out step by step. Calculate smartly when possible.

(a) 77 × 101 + 33

(b) 46 × 64 − 64 × 36

(c) 432 × 15 ÷ 20

(d) 3200 ÷ 40 ÷ 8

4 Write the number sentences and then calculate.

(a) A is 480 and it is 6 times B. What is the sum of A and B?

(b) The sum of 565 and 19 is divided by 50. What is the quotient?

5 Fill in the missing numbers.

(a) Split the same multiplication sentence in different ways.

25 × 36   25 × 36   25 × 36

= _____   = _____   = _____

= _____   = _____   = _____

= _____   = _____   = _____

(b) Subtract 80 from 425 for ☐ times until you get the remainder 25.

(c) When calculating 25 × 33, Hardeep mistook one of the two 3s for a different number. As a result, the product is 50 greater than the correct answer. Hardeep mistook 33 for ☐.

(d) 720 ÷ 60 = 720 ÷ 6 ÷ ☐ = ☐ ÷ ☐ = ☐

(e) When the quotient of 440 ÷ ■0 is a 2-digit number, the greatest possible number in the ■ is ☐.

Challenge and extension questions

6 Hanna mistook ▲ ÷ 50 for ▲ ÷ 500, and therefore the quotient she got was 12.

(a) What should the correct quotient be? ☐

(b) What number does ▲ stand for? ☐

7 A circular pond has a perimeter of 300 metres. If a willow tree is planted every 5 metres around the pond, how many willow trees are needed?

Chapter 3 test

1 Calculate mentally and then write the answers.

(a) $125 \times 4 =$ ____

(b) $111 \times 30 =$ ____

(c) $950 \div 20 =$ ____

(d) $300 \div 30 =$ ____

(e) $240 \times 40 =$ ____

(f) $8 \times 125 =$ ____

(g) $2500 \div 50 =$ ____

(h) $4900 \div 70 =$ ____

(i) $3 \times 125 =$ ____

(j) $121 \div 11 =$ ____

(k) $5 \times 5 \times 50 =$ ____

(l) $144 \div 12 \div 2 =$ ____

2 Use the column method to calculate. Check the answer to the last question in each row.

(a) $207 \times 43 =$

(b) $880 \times 2300 =$

(c) $492 \times 20 =$

(d) $416 \times 47 =$

(e) $4256 \div 30 =$

(f) $16\,919 \div 50 =$

3 Work these out step by step. Calculate smartly when possible.

(a) 2623 − 1746 + 1377

(b) 203 × 15 + 85

(c) 45 × 235 − 45 × 135

(d) 909 × 20 ÷ 10

(e) 2400 ÷ 20 ÷ 4

(f) 39 × (100 − 25)

4 Write the answers.

(a) The product of 8500 × 160 has ☐ zeros at the end.

(b) The sum of the smallest 3-digit number and the greatest 3-digit number is ☐ .

(c) The quotient of 5000 ÷ 50 is a ☐ -digit number. The highest value place of the quotient is in the _____ place. There are ☐ zeros at the end of the quotient.

(d) In $60\overline{)\ \blacksquare\ 2\ 6\ 5\ }$, if the quotient is a 3-digit

number, the smallest possible number in the $\blacksquare$ is [ ].

If the quotient is a 2-digit number, the number in the $\blacksquare$

could be _____ .

(e) To saw a 125-metre-long log into 5 pieces, it needs to be sawn
[ ] times.

5 Multiple choice questions. (For each question, choose the correct answer and write the letter in the box.)

(a) To calculate 31 × 29, the wrong method in the following is [ ]

A. 31 × 30 − 31 × 1   B. 31 × 20 + 31 × 9

C. 30 × 29 + 29   D. (31 − 1) × (29 + 1)

(b) 99 is divided by a number, and the quotient is a one-digit

number. The smallest possible number of the divisor is [ ]

A. 100   B. 99   C. 10   D. 11

(c) A 3-digit number is multiplied by a 2-digit number, and the

product is a [ ]-digit number.

A. six   B. five   C. five or four   D. not sure

(d) Joe and Sarai live in the same apartment building. Joe lives on

the fifth floor, and Sarai lives on the second floor. The number

of steps between neighbouring floors is the same. If Sarai

needs to walk 36 steps from the ground floor to her home,

then Joe needs to walk [ ] steps from the ground floor to

his home.

A. 72   B. 90   C. 108   D. 120

6 Application problems.

(a) In a supermarket, eggs are packed in cartons with each holding 12 eggs. A box contains 10 cartons of eggs. A canteen bought 12 boxes of eggs. How many eggs did it buy? If the canteen uses 80 eggs a day, how many days can these eggs last?

(b) A fruit shop received a delivery of 3680 kilograms of watermelon in the morning. This was half of the weight delivered in the afternoon. How many kilograms of watermelon did the shop receive in total on that day?

(c) The water pump on a building site pumped 730 tonnes of water during the day and 350 tonnes at night. If it works 20 hours in the whole day, how much water did it pump every hour on average?

(d) A shop purchased 100 pieces each of the normal Rubik's cubes and irregular Rubik's cubes as shown in the diagram. After selling out of both types of cube the shop had made a profit of £200 from the sale of the normal Rubik's cubes and £500 from the sale of the irregular Rubik's cubes. What price did the shop pay for each type of cube?

£10

£85

Chapter 4 Addition and subtraction of fractions

4.1 Fractions in hundredths

 Learning objective Recognise and use hundredths

 Basic questions

1 Write the missing fraction in each box.

(a) If a whole is divided into 2 equal parts,

each part is ☐ of the whole.

(b) If a whole is divided into 10 equal parts,

3 parts are ☐ of the whole.

(c) If a whole is divided into 100 equal parts,

each part is ☐ of the whole.

(d) If a whole is divided into 200 equal parts,

3 parts are ☐ of the whole.

2 Shade each diagram to represent the fraction given. The first one has been done for you.

(a) $\frac{1}{4}$ or $\frac{25}{100}$

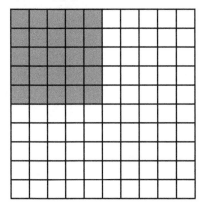

(b) $\frac{1}{10}$ or $\frac{10}{100}$

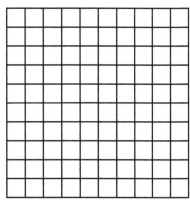

(c) $\frac{1}{100}$

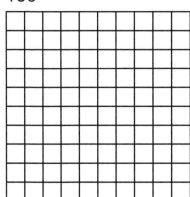

(d) $\frac{89}{100}$

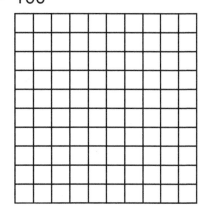

3 Circle equivalent fractions and draw lines to link them. One has been done for you.

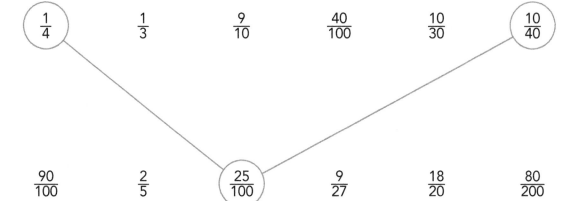

4 Mark the following fractions on the number line.

(a) $A = \frac{1}{100}$ $B = \frac{29}{100}$ $C = \frac{1}{10}$ $D = \frac{97}{100}$ $E = \frac{1}{2}$ $F = \frac{79}{100}$

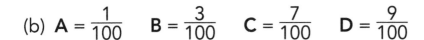

0   1 or $\frac{100}{100}$

(b) $A = \frac{1}{100}$ $B = \frac{3}{100}$ $C = \frac{7}{100}$ $D = \frac{9}{100}$

0 $\frac{10}{100}$

5 Count in hundredths and fill in the missing numbers.

(a) $\frac{1}{100}, \frac{3}{100}, \frac{5}{100},$ ☐ , ☐ , ☐ ,

(b) $\frac{1}{100}, \frac{11}{100}, \frac{21}{100}, \frac{31}{100},$ ☐ , ☐ , ☐ ,

(c) $\frac{97}{100}, \frac{93}{100}, \frac{89}{100},$ ☐ , ☐ , ☐ ,

6 A school bought 200 books and gave them to different classes. Write a suitable fraction in each box.

(a) Class A received 20 books. This is ☐ of the books bought.

(b) Class B received 38 books. This is ☐ of the books bought.

(c) Class C received 60 books. This is ☐ of the books bought.

(d) Class D received 42 books. This is ☐ of the books bought.

(e) All the remaining books were given to Class E. This is ☐ of the books bought.

Challenge and extension question

7 In a community fun run all the participants were divided into 10 equal groups. The participants in the first group were further divided into 10 equal subgroups. Answer the following:

(a) What fraction of all the participants was the number of the participants in each subgroup of the first group?

(b) If the number of all the participants was exactly 500, how many participants were there in each group? How many participants were there in each subgroup of the first group?

4.2 Addition and subtraction of fractions (1)

Learning objective Add fractions with the same denominator

Basic questions

1 Think carefully. Calculate and fill in the boxes.

(a) $\dfrac{3}{10} + \dfrac{4}{10} = \dfrac{3+4}{10} = \dfrac{\boxed{}}{10}$

We can also calculate this way: adding $\boxed{}$ lots of $\dfrac{1}{10}$ and $\boxed{}$ lots of $\dfrac{1}{10}$, the sum is $\boxed{}$ lots of $\dfrac{1}{10}$, which is $\boxed{}$.

We can represent the addition using the diagram below. (Can you represent this another way to represent this?)

$\dfrac{3}{10} + \dfrac{4}{10} = \dfrac{\boxed{}}{\boxed{}}$

| $\frac{1}{10}$ | $\frac{1}{10}$ | $\frac{1}{10}$ | $\frac{1}{10}$ | $\frac{1}{10}$ | $\frac{1}{10}$ | $\frac{1}{10}$ | | | |

$\underbrace{\qquad}_{\frac{3}{10}}$ $\underbrace{\qquad}_{\frac{4}{10}}$

(b) $\frac{7}{17} + \frac{9}{17} = \frac{7+9}{17} = \frac{\boxed{}}{17}$

We can also calculate this way: adding $\boxed{}$ lots of $\frac{1}{17}$

and $\boxed{}$ lots of $\frac{1}{17}$, the sum is $\boxed{}$ lots of $\frac{1}{17}$, which is $\boxed{}$.

Now draw a diagram to represent $\frac{7}{17} + \frac{9}{17}$.

(c) In general, adding fractions with the same denominator means

_____ the numerators and keeping the

denominators _____.

2 Calculate.

(a) $\frac{1}{7} + \frac{3}{7} = \boxed{}$

(b) $\frac{2}{5} + \frac{1}{5} = \boxed{}$

(c) $\frac{9}{20} + \frac{4}{20} = \boxed{}$

(d) $\frac{6}{43} + \frac{16}{43} = \boxed{}$

(e) $\frac{50}{77} + \frac{10}{77} = \boxed{}$

(f) $\frac{115}{800} + \frac{345}{800} = \boxed{}$

(g) $\frac{3}{9} + \frac{4}{9} + \frac{1}{9} = \boxed{}$

(h) $\frac{5}{32} + \frac{6}{32} + \frac{7}{32} + \frac{8}{32} = \boxed{}$

3 Fill in the boxes.

(a) ☐ lots of $\frac{1}{15}$ are $\frac{13}{15}$. 8 lots of $\frac{1}{9}$ are ☐.

(b) 5 lots of $\frac{1}{6}$ are ☐. Adding it to 1 lot of $\frac{1}{6}$ is ☐.

(c) In a box full of balls, 4 are red, which is $\frac{4}{11}$ of the whole box. If we add one more red ball, then the red balls are ☐ of the whole box.

4 Write the number sentences and then calculate.

(a) What is the sum of 5 lots of $\frac{1}{20}$ and 7 lots of $\frac{1}{20}$?

(b) After $\frac{3}{19}$ is taken away from a number, the result is $\frac{11}{19}$. Find the number.

5 In an auto rally, the Speed team drove $\frac{3}{10}$ of the whole journey on the first day. On the second day, they drove $\frac{4}{10}$ of the whole journey and on the third day they drove $\frac{2}{10}$ of the journey.

How much of the journey did the team drive in the first three days? Use a fraction to show your answer.

Challenge and extension questions

6 Write a suitable fraction or a number in each box.

(a) $\dfrac{3}{10} + \boxed{} = \dfrac{1}{2}$

(b) $\dfrac{3}{10} + \boxed{} = 1$

(c) $\dfrac{9}{22} + \boxed{} = \dfrac{1}{2}$

(d) $\dfrac{20}{24} = \dfrac{\boxed{}}{48} = \dfrac{5}{\boxed{}}$

7 Look for patterns and then fill in each $\boxed{}$ with a suitable number.

$\dfrac{1}{2} = \dfrac{1}{3} + \dfrac{1}{6}, \ \dfrac{1}{3} = \dfrac{1}{4} + \dfrac{1}{12}, \ \dfrac{1}{4} = \dfrac{1}{5} + \dfrac{1}{20}$

$\dfrac{1}{5} = \dfrac{1}{\boxed{}} + \dfrac{1}{\boxed{}}, \ \dfrac{1}{9} = \dfrac{1}{\boxed{}} + \dfrac{1}{\boxed{}}, \ \dfrac{1}{50} = \dfrac{1}{\boxed{}} + \dfrac{1}{\boxed{}}$

4.3 Addition and subtraction of fractions (2)

Learning objective Subtract fractions with the same denominator

Basic questions

1 Think carefully. Calculate and fill in the boxes.

(a) $\dfrac{9}{10} - \dfrac{3}{10} = \dfrac{9-3}{10} = \dfrac{\boxed{}}{10}$

We can also calculate this way: $\boxed{}$ lots of $\dfrac{1}{10}$ minus $\boxed{}$ lots of $\dfrac{1}{10}$. The difference is $\boxed{}$ lots of $\dfrac{1}{10}$, which is $\boxed{}$.

We can represent the subtraction using the diagram below. (Can you represent this another way?)

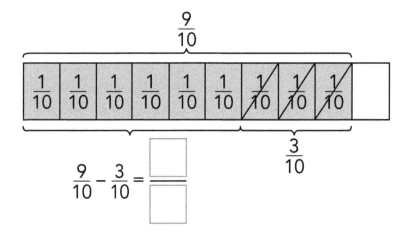

$\dfrac{9}{10} - \dfrac{3}{10} = \dfrac{\boxed{}}{\boxed{}}$

(b) $\frac{16}{17} - \frac{9}{17} = \frac{16-9}{17} = \frac{\boxed{}}{17}$

We can also calculate this way: $\boxed{}$

lots of $\frac{1}{17}$ minus $\boxed{}$ lots of $\frac{1}{17}$, the difference is $\boxed{}$.

Now draw a diagram to represent $\frac{16}{17} - \frac{9}{17}$.

(c) In general, subtracting fractions with the same denominator

means _____ the numerators and keeping the

denominators _____.

2 Calculate.

(a) $\frac{5}{6} - \frac{4}{6} = \boxed{}$

(b) $\frac{9}{14} - \frac{3}{14} = \boxed{}$

(c) $\frac{17}{30} - \frac{8}{30} = \boxed{}$

(d) $\frac{29}{72} - \frac{16}{72} = \boxed{}$

(e) $\frac{90}{300} - \frac{40}{300} = \boxed{}$

(f) $\frac{8}{25} - \frac{6}{25} = \boxed{}$

(g) $1 - \frac{8}{65} - \frac{9}{65} - \frac{10}{65} = \boxed{}$

(h) $\frac{\boxed{}}{36} - \frac{10}{36} = \frac{11}{36}$

(i) $\frac{10}{100} - \frac{\boxed{}}{100} = \frac{3}{100}$

3 Write the number sentences and then calculate.

(a) After taking away $\frac{3}{15}$ from a minuend, the result is $\frac{11}{15}$. Find the minuend.

(b) What is the result of taking away 4 lots of $\frac{1}{20}$ from $\frac{8}{20}$?

(c) What is the difference when subtracting 3 lots of $\frac{1}{7}$ from 1?

Challenge and extension questions

4 Let's calculate.

(a)
$$\frac{4}{9} + \frac{5}{9} - \frac{9}{10}$$

(b)
$$1 - \frac{10}{10} + \frac{9}{10}$$

(c)
$$\frac{7}{12} + \frac{5}{9} + \frac{5}{12} + \frac{4}{9}$$

(d)
$$1 - \frac{1}{5} - \frac{3}{8} + \frac{1}{5}$$

5 At Marnie birthday, Dad gave $\frac{3}{4}$ of the cake to Grandpa, Grandma, Aunty and Cousin Theo. Marnie and Mum had $\frac{1}{8}$ of the cake each. Was there any cake left for Dad himself?

4.4 Fun with exploration – 'fraction wall'

Learning objective Compare, add and subtract fractions

Basic questions

1 Learning buddy – 'fraction wall'.
Look at the fraction wall below. What do you notice?

1															
$\frac{1}{2}$								$\frac{1}{2}$							
$\frac{1}{3}$					$\frac{1}{3}$						$\frac{1}{3}$				
$\frac{1}{4}$				$\frac{1}{4}$				$\frac{1}{4}$				$\frac{1}{4}$			
$\frac{1}{5}$			$\frac{1}{5}$			$\frac{1}{5}$			$\frac{1}{5}$			$\frac{1}{5}$			
$\frac{1}{6}$		$\frac{1}{6}$		$\frac{1}{6}$		$\frac{1}{6}$		$\frac{1}{6}$		$\frac{1}{6}$					
$\frac{1}{7}$		$\frac{1}{7}$		$\frac{1}{7}$		$\frac{1}{7}$		$\frac{1}{7}$		$\frac{1}{7}$		$\frac{1}{7}$			
$\frac{1}{8}$		$\frac{1}{8}$		$\frac{1}{8}$		$\frac{1}{8}$		$\frac{1}{8}$		$\frac{1}{8}$		$\frac{1}{8}$		$\frac{1}{8}$	
$\frac{1}{9}$	$\frac{1}{9}$		$\frac{1}{9}$		$\frac{1}{9}$		$\frac{1}{9}$		$\frac{1}{9}$		$\frac{1}{9}$		$\frac{1}{9}$		$\frac{1}{9}$
$\frac{1}{10}$	$\frac{1}{10}$	$\frac{1}{10}$	$\frac{1}{10}$	$\frac{1}{10}$	$\frac{1}{10}$	$\frac{1}{10}$	$\frac{1}{10}$	$\frac{1}{10}$	$\frac{1}{10}$						
$\frac{1}{12}$	$\frac{1}{12}$	$\frac{1}{12}$	$\frac{1}{12}$	$\frac{1}{12}$	$\frac{1}{12}$	$\frac{1}{12}$	$\frac{1}{12}$	$\frac{1}{12}$	$\frac{1}{12}$	$\frac{1}{12}$	$\frac{1}{12}$				
$\frac{1}{16}$	$\frac{1}{16}$	$\frac{1}{16}$	$\frac{1}{16}$	$\frac{1}{16}$	$\frac{1}{16}$	$\frac{1}{16}$	$\frac{1}{16}$	$\frac{1}{16}$	$\frac{1}{16}$	$\frac{1}{16}$	$\frac{1}{16}$	$\frac{1}{16}$	$\frac{1}{16}$	$\frac{1}{16}$	$\frac{1}{16}$

The fraction wall is very helpful for comparing fractions with the same denominators or the same numerators. It makes it much easier to see the two fractions being compared, making the comparison straightforward.

A fraction wall can also help us visualise the addition and subtraction of fractions with the same denominators.

Drawing a vertical line from the top to the bottom in a fraction wall can help us quickly find the equal fractions.

2 Use the fraction wall to compare the fractions. Then write >, < or = in each circle.

(a) $\frac{5}{12}$ ◯ $\frac{8}{12}$ (b) $\frac{7}{9}$ ◯ $\frac{4}{9}$ (c) $\frac{1}{10}$ ◯ $\frac{3}{10}$

(d) $\frac{4}{8}$ ◯ $\frac{1}{2}$ (e) $\frac{7}{16}$ ◯ $\frac{7}{10}$ (f) $\frac{10}{12}$ ◯ $\frac{10}{16}$

3 Do the calculation first and then check your answers using the fraction wall shown in Question 1.

(a) $\frac{1}{9} + \frac{8}{9} =$ ☐   (b) $\frac{11}{12} - \frac{9}{12} =$ ☐   (c) $\frac{1}{7} + \frac{2}{7} + \frac{3}{7} =$ ☐

4 Find $\frac{4}{6}, \frac{2}{3}, \frac{3}{4}$ and $\frac{1}{4}$ in the fraction shown in Question 1.

(a) The fractions that are same as $\frac{4}{6}$ are _____.

(b) The fractions that are same as $\frac{2}{3}$ are _____.

(c) The fractions that are same as $\frac{3}{4}$ are _____.

(d) The fractions that are same as $\frac{1}{4}$ are _____.

Challenge and extension question

5 Fun with the subtraction of fractions.

Do the following: Take away half $\left(\frac{1}{2}\right)$ from 1, and then take away half of the answer $\left(\frac{1}{4}\right)$ from $\frac{1}{2}$ and again take away half of the answer $\left(\frac{1}{8}\right)$ from $\frac{1}{4}$ and then continue to take away half of the answer $\left(\frac{1}{16}\right)$ from $\frac{1}{8}$...

The calculation is a bit challenging, but the result shows something interesting.

Try it on your own and write the result.

$$1 - \frac{1}{2} - \frac{1}{4} - \frac{1}{8} - \frac{1}{16} = \boxed{}$$

Chapter 4 test

1 Calculate mentally and then write the answers.

(a) $7 \times 11 + 7 = \boxed{}$

(b) $56 \div 8 \times 9 = \boxed{}$

(c) $60 \div 30 + 34 = \boxed{}$

(d) $200 \div 5 \times 8 = \boxed{}$

(e) $12 \times 50 - 100 = \boxed{}$

(f) $180 - 99 + 1 = \boxed{}$

(g) $\frac{2}{7} + \frac{1}{7} = \boxed{}$

(h) $\frac{6}{23} + \frac{17}{23} = \boxed{}$

(i) $\boxed{} + \frac{1}{16} = \frac{9}{16}$

(j) $1 - \frac{7}{13} = \boxed{}$

(k) $\frac{4}{5} - \frac{1}{5} = \boxed{}$

(l) $\boxed{} - \frac{3}{16} = \frac{1}{2}$

2 Work these out step by step.

(a) $\frac{1}{8} + \frac{3}{8} + \frac{2}{8}$

(b) $\frac{10}{11} - \frac{7}{11} - \frac{1}{11}$

(c) $\frac{3}{14} + \frac{5}{14} - \frac{4}{14}$

(d) $\frac{12}{25} - \frac{4}{25} + \frac{9}{25}$

(e) $\frac{7}{9} + \frac{2}{9} - \frac{31}{74}$

(f) $\frac{5}{14} + \frac{6}{14} - \frac{7}{14}$

3 Shade each diagram to represent the fraction given.

(a)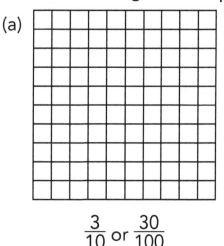

$$\frac{3}{10} \text{ or } \frac{30}{100}$$

(b)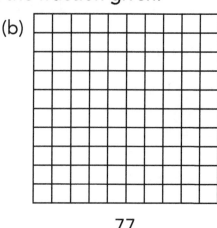

$$\frac{77}{100}$$

4 Fill in the boxes.

(a) $\frac{3}{8} = \frac{\boxed{}}{16} = \frac{12}{\boxed{}}$

(b) $\frac{1}{5} = \frac{\boxed{}}{10} = \frac{\boxed{}}{100}$

(c) $\boxed{}$ lots of $\frac{1}{5}$ makes 1.

(d) 50 lots of $\frac{1}{100}$ makes $\boxed{}$

(e) $\boxed{}$ lots of $\boxed{}$ is $\frac{7}{11}$.

(f) $\frac{1}{10}$ divided by 10 makes $\boxed{}$.

(g) Adding $\boxed{}$ lots of $\frac{1}{18}$ to $\frac{13}{18}$, the result equals 1.

Taking away $\boxed{}$ lots of $\frac{1}{18}$ from $\frac{13}{18}$, the result equals $\frac{1}{2}$.

5 Multiple choice questions. (For each question, choose the correct answer and write the letter in the box.)

(a) Fold a rope in half twice and then fold it in half one more time. Now each part is $\boxed{}$ of the whole.

A. $= \frac{1}{2}$ B. $= \frac{1}{4}$ C. $= \frac{1}{8}$ D. $= \frac{1}{16}$

(b) Read the statements below. Statement ☐ is incorrect.

A. Dividing a square into 100 equal parts, each part is $\frac{1}{100}$ of the whole.

B. Dividing one tenth of a square by 100, each part is $\frac{1}{100}$ of the whole.

C. Dividing one tenth of a square by 10, each part is $\frac{1}{100}$ of the whole.

D. Dividing half of a square by 50, each part is $\frac{1}{100}$ of the whole.

6 Write the number sentences and then calculate.

(a) How much greater is $\frac{4}{5}$ than $\frac{1}{5}$?

(b) What is the result of $\frac{79}{80}$ minus $\frac{50}{80}$ and then plus $\frac{30}{80}$?

7 Application problems.

(a) The bookshop is delivering some new books to school for the pupils. A van can load 1200 books per trip. How many books can be delivered by four vans in two trips?

Answer: _____

(b) A bundle of pencils was shared among four children. Bella got 4 pencils, Maya got 5 pencils, Imran got 6 pencils and Jordan got 7 pencils. What fraction of the total pencils did Maya get?

Answer: _____

(c) A box of tea weighs 1 kilogram. How many grams of tea were left after $\frac{1}{4}$ of the tea was taken from the box?

Answer: _____

Chapter 5 Consolidation and enhancement

5.1 Multiplication and multiplication tables

Learning objective Recall multiplication facts and solve multiplication problems

Basic questions

1 Complete the multiplication table.

×	1	2	3	4	5	6	7	8	9	10	11	12
1		2										
2			6									
3				12								
4					20							
5						30						
6							42					
7								56				
8									72			
9										90		
10											110	
11												132
12	12											

2 Complete the multiplication facts. Then write two related multiplication sentences and two related division sentences.

(a) 7 times 8 is _____ .

(b) 8 times 11 is _____ .

(c) 5 times 12 is _____ .

(d) 6 times _____ is 60.

(e) _____ times 9 is 72.

(f) _____ times 12 is 132.

3 Represent the following repeated additions as multiplications and then write the answers.

(a) $3 + 3 + 3 + 3 = \boxed{} \times \boxed{} = \boxed{}$

(b) $6 + 6 + 6 + 6 + 6 + 6 + 6 + 6 + 6 = \boxed{} \times \boxed{} = \boxed{}$

(c) $11 + 11 + 11 + 11 + 11 + 11 = \boxed{} \times \boxed{} = \boxed{}$

(d) $7 + 7 + 7 + 7 + 7 = \boxed{} \times \boxed{} = \boxed{}$

(e) $12 + 12 + 12 + 12 + 12 + 12 + 12 + 12 = \boxed{} \times \boxed{} = \boxed{}$

4 A new school enrolled 40 pupils in its first year. During the second year, enrolment doubled, and in the third year it tripled. What was the enrolment in the second year? What was the enrolment in the third year? Write the number sentences and find the answers.

5 Solve the following multiplication problems.

(a) There are 3 T-shirts and 4 pairs of trousers. Each T-shirt can be matched with any of the trousers. How many different outfits can you make with these T-shirts and trousers?

(b) A cafe serves four choices of burger – beef, chicken, fish and vegetarian – and five choices of drink – juice, coffee, cola, water and hot chocolate. Erin wants one burger and one drink. How many different combinations can she have?

Challenge and extension questions

6 There are two roads from Town A to Town B, three roads from Town B to Town C and two roads from Town C to Town D. How many different ways are there to travel by road from Town A to Town D, passing through Town B and then Town C?

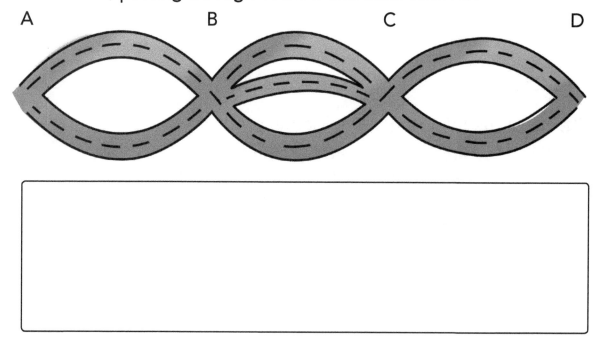

7 Two chess pieces are to be placed in two different squares on the grid paper on the right. They cannot be placed in either the same row or the same column. How many different ways can these two chess pieces be placed?

5.2 Relationship between addition and subtraction

Learning objective Use the inverse relationship between addition and subtraction

Basic questions

1 Fill in the boxes.

(a) (i) 50 + 80 = ☐

☐ + 80 = 130

130 − 80 = ☐

(ii) 53 + 47 = ☐

47 + ☐ = 100

100 − ☐ = ☐

(iii) 230 − 70 = ☐

70 + ☐ = 230

☐ − 160 = 70

(iv) 190 − 90 = ☐

190 − ☐ = 90

☐ − 90 = ☐ + 10

(b) Addend = _____ − addend

Minuend − subtrahend = _____

Subtrahend = minuend − _____

_____ = difference + subtrahend

Subtraction is the inverse operation of _____.

2 Find the number that goes in the ■.

Show your working. The first one has been done for you.

(a) 165 − ■ = 88

> ■ = 165 − 88
>
> ■ = 77

(b) 76 + ■ = 311

(c) ■ + 901 = 1002

(d) ■ − 190 = 75

3 Use the relationship between addition and subtraction to complete the following number sentences.

(a) 419 + 363 = 782

782 ◯ 363 = 419

782 ◯ 419 = 363

(b) 1950 − 1016 = 934

1950 ◯ 934 = 1016

1016 ◯ 934 = 1950

(c) 756 + 112 = 868

112 = _____

756 = _____

(d) ■ − ● = ▲

■ = _____

● = _____

4 Write the number sentences and then calculate.

(a) An addend is 126. The sum of this addend and another addend is 789. Find the other addend.

Answer: _____

(b) A minuend is 120 and the difference is half of the minuend. Find the subtrahend.

Answer: _____

Challenge and extension questions

5 True or false? (Put a ✓ for true and a ✗ for false in each box.)

(a) If A − 139 = 1080, then A = 1080 − 139. . . . . ☐

(b) In a subtraction sentence, if the subtrahend equals the difference, the minuend must be twice the difference. . . . . . . ☐

6 Tom made a mistake when he was working on an addition sentence. He accidentally wrote an addend, 36, as 63 and the sum he got was 278. What was the other addend?

Answer: _____

5.3 The relationship between multiplication and division

Learning objective Use the inverse relationship between multiplication and division

Basic questions

1 Fill in the boxes.

(a) (i) $22 \times 5 = \boxed{}$

   $\boxed{} \times 5 = 110$

   $110 \div \boxed{} = 5$

(ii) $120 \div 10 = \boxed{}$

   $12 \times 10 = \boxed{}$

   $120 = \boxed{} \times 10$

(iii) $125 \times 8 = \boxed{}$

   $1000 = \boxed{} \times 125$

   $\boxed{} \times 8 = 1000$

(iv) $35 \times 4 = \boxed{}$

   $4 \times \boxed{} = 140$

   $140 \div 4 = \boxed{}$

(b) Factor = _____ ÷ factor

   _____ = quotient × divisor

   Dividend ÷ _____ = divisor

(c) Choose the correct answer and put a ✓ in the box.

   The product of multiplication is equivalent to which term of its related division?

   dividend $\boxed{}$ divisor $\boxed{}$ quotient $\boxed{}$

2 Find the number that goes in the ■. Show your working. The first one has been done for you.

(a) $15 \times ■ = 150$

$$■ = 150 \div 15$$
$$= 10$$

(b) $960 \div ■ = 80$

(c) $■ \div 9 = 12$

(d) $■ \times 70 = 1050$

(e) $■ \times 21 = 189$

(f) $2800 - ■ = 40$

3 Write the number sentences and then calculate.

(a) A factor is 8. When multiplied by another factor, the product is 768. Find the other factor.

Number sentence: _____

Answer: _____

(b) A dividend is 288. The quotient and the remainder are 2 and 4 respectively. Find the divisor.

Number sentence: _____

Answer: _____

(c) 840 is 30 times a number. What is the number?

Number sentence: _____

Answer: _____

Challenge and extension questions

4 When Lily was multiplying two numbers, she misread a factor 32 as 30, and got a product of 360. Think carefully and find the correct product.

Answer: _____

5 A number can be exactly divided by another number and the quotient is 9. The sum of the dividend and divisor is 210. What are the dividend and divisor?

Answer: _____

5.4 Multiplication by 2-digit numbers

Learning objective Use written methods to multiply by 2-digit numbers

Basic questions

1 Calculate mentally and then write the answers.

(a) 220 × 4 = [ ]

(b) 900 ÷ 20 = [ ]

(c) 25 × 30 = [ ]

(d) 1200 ÷ 40 = [ ]

(e) 8100 ÷ 90 = [ ]

(f) 102 × 20 = [ ]

(g) 60 ÷ 30 = [ ]

(h) 700 ÷ 35 = [ ]

(i) 30 × 12 = [ ]

(j) 220 ÷ 11 = [ ]

(k) 42 × 4 = [ ]

(l) 800 ÷ 50 = [ ]

2 Use the column method to calculate. Check the answers to the questions marked with *.

(a) 48 × 126 = [ ]

(b) *2610 ÷ 30 = [ ]

(c) *270 × 60 = [ ]

3 Use the four digits 8, 4, 2 and 5 to make a multiplication of two 2-digit numbers.

(a) What is the greatest possible product?

(b) What is the smallest possible product?

4 In 2014, Meera's family saved £25 per month on their water bill and £19 per month on their electricity bill. How much did they save altogether on their water and electricity bills for the whole year?

Answer: _____

5 A school paid £9720 for electricity for the whole of 2014. In response to an energy-saving campaign, the school launched an 'electricity saving plan' in 2015, aiming to save £90 per month compared to 2014. According to this plan, how much would the school pay for electricity in 2015?

Answer: _____

Challenge and extension questions

6 Fill in the missing numbers.

(a) In multiplying 914 × 62, the product of the digit 9 and the digit 6 is the same as ☐ × ☐ .

(b) The product of 27 × 42 is a ☐ -digit number. It is between ☐ and ☐ , nearer to ☐ .

(c) There are ☐ zeros at the end of the product of 25 × 800.

7 Fill in the boxes to complete the calculation.

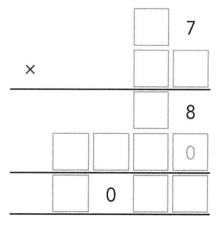

5.5 Practice with fractions

Learning objective Calculate fractions of amounts and add and subtract fractions

Basic questions

1 Use fractions to represent the shaded parts.

(a) 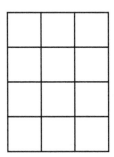   (b) 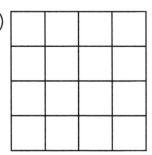   (c)

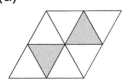   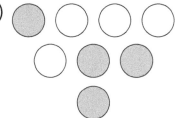   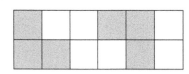

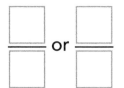   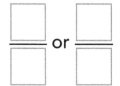

2 Colour $\frac{1}{4}$ of each grid.

(a)   (b)   (c)

3 Fill in the answers.

(a) When you divide a pizza into 12 equal pieces, each piece is

☐ of the whole.

(b) There are ☐ lots of $\frac{1}{6}$ in 1.

(c) The denominator is 23 and it is 6 greater than the numerator.

The fraction is ☐.

(d) 5 lots of $\frac{1}{6}$ is ☐. 8 lots of $\frac{1}{9}$ is ☐.

(e) $\frac{7}{11}$ consists of ☐ lots of $\frac{1}{11}$.

(f) ☐ lots of $\frac{1}{4}$ equals 1.

(g) Ava has 8 sweets. If she eats $\frac{1}{4}$ of them, that means that she has eaten ☐ sweets.

(h) $\frac{1}{2}$ of 18 is ☐. $\frac{2}{3}$ of 12 is ☐.

(i) There are _____ fractions that are equal to $\frac{2}{3}$.

(j) 4 lots of $\frac{1}{8}$ is equal to ☐ lots of $\frac{1}{10}$.

4 Calculate.

(a) $\frac{1}{4} + \frac{2}{4} =$ ☐

(b) $\frac{118}{250} - \frac{50}{250} =$ ☐

(c) $\frac{11}{50} - \frac{7}{50} + \frac{8}{50} =$ ☐

(d) $1 - \frac{5}{7} + \frac{4}{7} =$ ☐

5 Put these fractions in order, starting with the greatest.

$\boxed{\frac{99}{100}}$ $\boxed{\frac{1}{100}}$ $\boxed{\frac{1}{10}}$ $\boxed{\frac{1}{2}}$ $\boxed{\frac{49}{100}}$

☐ ☐ ☐ ☐ ☐

6 Think carefully and then fill in the boxes.

(a) A 1-metre-long ribbon is cut into 8 equal pieces.

The length of each piece is ☐ of 1 metre.

This is ☐ of a metre long.

Five pieces together are ☐ of a metre long.

(b) $1 - \dfrac{\boxed{}}{7} + \dfrac{2}{7} = \dfrac{4}{7}$

(c) $\boxed{} - \dfrac{8}{20} = \dfrac{1}{2}$

(d) $\dfrac{2}{3} = \dfrac{\boxed{}}{6} = \dfrac{12}{\boxed{}} = \dfrac{18}{\boxed{}} = \dfrac{\boxed{}}{12} = \dfrac{\boxed{}}{24}$

5.6 Roman numerals to 100

Learning objective Read Roman numerals for numbers to 100

Basic questions

1 These clocks have Roman numerals on them. Read the times and write them to the nearest minute in the spaces. Use the 12-hour format.

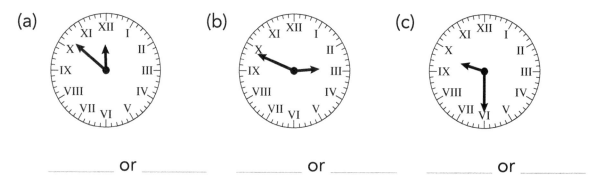

(a)

(b)

(c)

_____ or _____   _____ or _____   _____ or _____

2 Complete the table, writing the numbers from 1 to 100 in Roman numerals.

I				V					X
XI			XIV						XX
XXI					XXVI				XXX
XXXI		XXXIII							XL
XLI							XLVIII		L
LI				LV					LX
LXI								LXIX	LXX
LXXI					LXXVI				LXXX
LXXXI	LXXXII								XC
XCI						XCVII			C

3 Draw lines to match the Roman numerals to the numbers in digits.

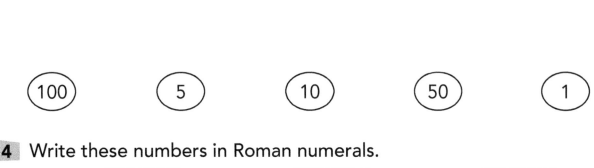

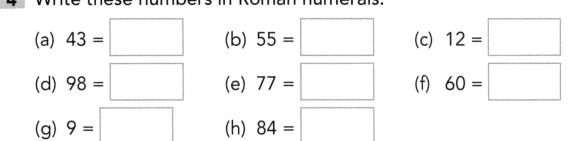

4 Write these numbers in Roman numerals.

(a) 43 = [ ]   (b) 55 = [ ]   (c) 12 = [ ]

(d) 98 = [ ]   (e) 77 = [ ]   (f) 60 = [ ]

(g) 9 = [ ]   (h) 84 = [ ]

5 Write these Roman numerals in digits.

(a) VI = [ ]   (b) LII = [ ]   (c) LXXIX = [ ]

(d) IX = [ ]   (e) XCV = [ ]   (f) LXV = [ ]

(g) XIX = [ ]   (h) LXX = [ ]

6 Sam was in the library looking at a very old book. The book had three chapters and the page numbers were written using Roman numerals starting from I. The first chapter ended at XXIV, the second chapter ended at LIV, and the third chapter ended at LX.

(a) Write the number of pages in each chapter in digits.

Chapter 1 [ ]   Chapter 2 [ ]   Chapter 3 [ ]

(b) How many pages were there in total? [ ]

Challenge and extension question

7 True or false? (Put a ✓ for true and a ✗ for false in each box.)

(a) There is no Roman numeral symbol for the number 0. . . . . ☐

(b) Roman numerals cannot represent numbers larger than 100.. . ☐

(c) Some Roman numerals are still used today. . . . . ☐

Chapter 5 test

1 Calculate mentally and then write the answers.

(a) $77 - 6 - 14 =$ ⬚

(b) $100 \div 20 + 125 =$ ⬚

(c) $6 \times 3 \times 10 =$ ⬚

(d) $44 \times 9 + 44 =$ ⬚

(e) $12 \times 60 =$ ⬚

(f) $6 \times 7 + 33 =$ ⬚

(g) $90 \div 3 - 17 =$ ⬚

(h) $35 - 35 \div 7 =$ ⬚

(i) $24 \div 6 \times 8 =$ ⬚

2 Fill in the boxes.

(a) ⬚ $+ 693 = 8000$

(b) ⬚ $- 147 = 592$

(c) $82 \times$ ⬚ $= 738$

(d) $1650 \div$ ⬚ $= 30$

(e) ⬚ $\div 103 = 13$

(f) $128 +$ ⬚ $= 256 - 126$

3 Use the column method to calculate.

(a) $39 \times 102 =$ ⬚

(b) $385 \times 68 =$ ⬚

(c) $1080 \div 40 =$ ⬚

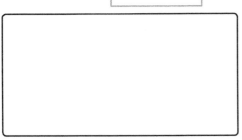

4 Work these out step by step.

(a) 234 × 55 ÷ 30

(b) 36 × 36 + 36 × 64

(c) (2012 − 162) ÷ 50

(d) 398 − 173 − 27 + 112

5 Calculate.

(a) $\dfrac{11}{15} - \dfrac{4}{15} - \dfrac{7}{15}$

(b) $\dfrac{26}{35} + \dfrac{9}{35} - \dfrac{11}{35}$

(c) $\dfrac{1}{100} + \dfrac{99}{100} - \dfrac{7}{9}$

6 Write the following Roman numerals in digits.

(a) VI =

(b) XIV =

(c) XCVII =

(d) XLII =

(e) LV =

(f) LXXIX =

7 Fill in the answers.

(a) When an addition is changed to a subtraction as

its _____ operation, the two addends are

equivalent to the _____ and _____,

and the sum is equivalent to the _____ in

the subtraction.

(b) If the quotient of 960 divided by a number is 30,

the number is ☐ .

(c) When a number is divided by 80, the quotient is 47.

The number is ☐ .

(d) When ☐ is subtracted from the greatest 4-digit

number, the result is the greatest 3-digit number.

(e) When Lily was doing a division, she mistakenly missed out the zero in the divisor 60, and therefore the quotient she got was 40. The correct quotient should be ☐ .

(f) If the difference of 3 ■ 7 − 257 is a 2-digit number, then the greatest possible number in the ■ must be ☐ .

If the difference is a 3-digit number, then the smallest possible number in the ■ must be ☐ .

8 Write the number sentences and then calculate.

(a) When a number is subtracted from 220 the difference is 96. Find the number.

Number sentence: _____

Answer: _____

(b) 5 eights divided by a number is 4. Find the number.

Number sentence: _____

Answer: _____

(c) If the difference of two numbers is 408, and the subtrahend is 65, what number is twice the minuend?

Number sentence: _____

Answer: _____

9 Application problems.

(a) Joe is reading a storybook. He has read 42 pages. The number of pages he has not read yet is 5 times the number he has read. How many pages has he not read yet?

Answer: _____

(b) The sum of the ages of Molly's mother and grandmother is 100 years. Molly's mother is 38 years old. Molly is 53 years younger than her grandmother. How old are Molly and her grandmother?

Answer: _____

(c) In the first half of a year, a school saved £98 per month on their water bill and £606 in total for the second half of the year. How much did the school save for the water bill in total in the whole year?

Answer: _____

(d) A school bought 36 tennis balls. This was four times the number of footballs the school bought. What is the difference between the number of tennis balls and the number of footballs?

Answer: _____

(e) Ming mistakenly wrote the minuend 4 in the ones place as 9, and 0 in the tens place as 6. He got a difference of 288. What is the correct difference?

Answer: _____

(f) Ellis has a green shirt, a blue shirt and a white shirt. He also has a pair of leather shoes and a pair of running shoes. In how many different combinations can Ellis wear his shirts and shoes?

Answer: _____

Chapter 6 Introduction to decimals

6.1 Decimals in life

 Learning objective Recognise decimal numbers in the context of money and measures

 Basic questions

1 Read the following information and then fill in the missing numbers.

In a week in July 2015, the price of unleaded petrol in a petrol station was 105.7p per litre and the price of diesel was 112.7p per litre.

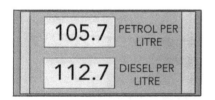

| 105.7 | PETROL PER LITRE |
| 112.7 | DIESEL PER LITRE |

Cola
1.8 litres

The volume of cola in a bottle sold in a supermarket is 1.8 litres.

PEANUTS

12.75 kg

The weight of a bag of peanuts for wild birds is 12.75 kg.

According to Guinness World Records, the tallest human in recorded history was Robert Wadlow (1918–1940), who lived in the United States. He reached 2.72 m in height.

(a) The decimal numbers used in the real-life contexts are

105.7, ☐ , ☐ , ☐ and ☐ .

Fill in the boxes below with whole numbers.

(b) The price of unleaded petrol was between ☐ pence and ☐ pence per litre, closer to ☐ pence per litre.

The price of diesel was between ☐ pence and ☐ pence per litre, closer to ☐ pence per litre.

(c) The volume of cola is between ☐ litres and ☐ litres. It is closer to ☐ litres.

(d) The weight of peanuts in the bag is between ☐ kg and ☐ kg. It is closer to ☐ kg.

(e) The height Robert Wadlow reached was between ☐ m and ☐ m. It is closer to ☐ m.

2 Read the following prices and fill in the boxes.

£13.20   £7.00   £21.68

(a) £13.20 is ☐ pounds and ☐ pence.

(b) £7.00 is ☐ pounds and ☐ pence.

(c) £21.68 is ☐ pounds and ☐ pence.

Challenge and extension question

3 Zarah and Ben had their weight and height measured in the school.

Zarah said, 'I saw the teacher writing down three numbers 1, 5 and 2 after he measured my height.'

What was Zarah's height? [   ]

Ben said, 'I saw the teacher writing down three numbers 3, 4 and 5 to record my weight.'

What was Ben's weight? [   ]

6.2 Understanding decimals (1)

Learning objective Recognise and write decimal equivalents to tenths, hundredths, thousandths, halves and quarters

Basic questions

1 Fill in the answers.

(a)

The shaded part is ☐ of the whole.

The unshaded part is ☐ of the whole.

The shaded part and the unshaded part added together

make a _____ .

(b)

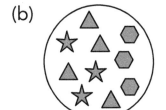

⭐ are ☐ of the whole.

▲ are ☐ of the whole.

⬡ are ☐ of the whole.

(c) Compare the fractions. Write >, < or = in each ◯.

$\frac{3}{10}$ ◯ $\frac{7}{10}$ $\qquad$ $\frac{35}{100}$ ◯ $\frac{3}{100}$ $\qquad$ $\frac{303}{1000}$ ◯ $\frac{303}{1000}$

(d) Fill in the boxes. Write a decimal to match the fraction or a fraction to match the decimal.

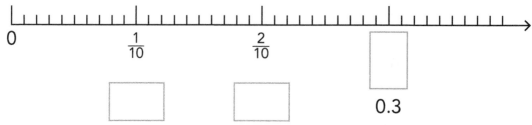

0.3

(e) All fractions with denominators of 10, 100, 1000 and so on can be expressed as _____.

$\frac{1}{10}$ can be written as a decimal number ☐,

it is read as _____.

$\frac{1}{100}$ can be written as ☐,

it is read as _____.

$\frac{1}{1000}$ can be written as ☐,

and it is read as _____.

2 Mark the fractions $\frac{1}{4}$, $\frac{1}{2}$ and $\frac{3}{4}$ on the number line and then write their decimal equivalents.

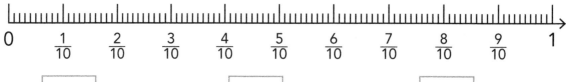

$\frac{1}{4}$ = ☐ $\qquad$ $\frac{1}{2}$ = ☐ $\qquad$ $\frac{3}{4}$ = ☐

◯ ◯ ◯ ◯ ◯ ◯ ◯ ○

3 Use decimal numbers to express the fractions and read them out.

(a) $\frac{7}{10}$ = ☐

    Read as: _____ .

(b) $\frac{16}{100}$ = ☐

    Read as: _____ .

(c) $\frac{256}{1000}$ = ☐

    Read as: _____ .

(d) $\frac{8}{10}$ = ☐

    Read as: _____ .

(e) $\frac{205}{1000}$ = ☐

    Read as: _____ .

(f) $\frac{95}{100}$ = ☐

    Read as: _____ .

4 Write the following decimals as fractions.

(a) 0.5 = ☐   (b) 0.03 = ☐   (c) 0.24 = ☐

(d) 0.001 = ☐   (e) 0.207 = ☐   (f) 0.9 = ☐

Challenge and extension questions

5 Fill in the boxes. The first one has been done for you.

(a) $\frac{3}{10} + \frac{4}{10} = \boxed{\frac{7}{10}} = \boxed{0.7}$

It is read as: <u>zero point seven</u>

(b) $\frac{35}{100} + \frac{27}{100} = \boxed{} = \boxed{}$

It is read as: _____

(c) $\frac{303}{1000} - \frac{28}{1000} = \boxed{} = \boxed{}$

It is read as: _____

(d) $\frac{24}{100} + \frac{19}{100} - \frac{13}{100} = \boxed{} = \boxed{}$

It is read as: _____

6 Shafiq and Lily like to drink a brand of milk sold in 200 ml cartons. Shafiq drank $\frac{4}{10}$ of a carton and Lily drank $\frac{5}{10}$ of another carton. Who has more milk left? How many millilitres did Shafiq drink? How many millilitres did Lily drink?

7 Number A and Number B are equal. Number A is $\frac{4}{5}$. The denominator of Number B is 10 greater than the denominator of Number A. What is the numerator of Number B?

6.3 Understanding decimals (2)

Learning objective Recognise and write decimal equivalents to tenths, hundredths and thousandths

Basic questions

1 Complete the place value chart to represent the decimal number 6735.482.

Whole number part				Decimal point	Decimal part		
Thousands place	Hundreds place	Tens place	Ones place	•	Tenths place	Hundredths place	Thousandths place
		3		•		8	

2 Count in decimals and complete the patterns.

(a) Count in 0.1s:

0, 0.1, 0.2, 0.3, ☐ , ☐ , ☐ , ☐ , ☐ , ☐ , 1.

(b) Count in 0.01s:

5.10, 5.11, 5.12, 5.13, ☐ , ☐ , ☐ , ☐ , ☐ ,

☐ , 5.20.

(c) Count back in 0.5s:

10, 9.5, 9, 8.5, ☐ , ☐ , ☐ , ☐ , ☐ , ☐ , 5.

3 Think carefully and fill in the boxes.

(a)

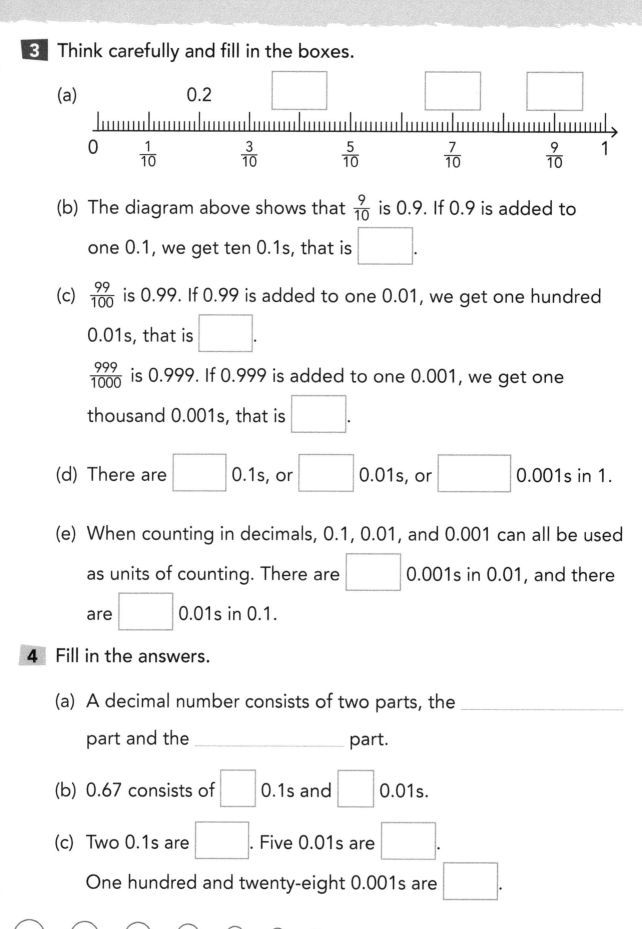

(b) The diagram above shows that $\frac{9}{10}$ is 0.9. If 0.9 is added to one 0.1, we get ten 0.1s, that is [ ].

(c) $\frac{99}{100}$ is 0.99. If 0.99 is added to one 0.01, we get one hundred 0.01s, that is [ ].

$\frac{999}{1000}$ is 0.999. If 0.999 is added to one 0.001, we get one thousand 0.001s, that is [ ].

(d) There are [ ] 0.1s, or [ ] 0.01s, or [ ] 0.001s in 1.

(e) When counting in decimals, 0.1, 0.01, and 0.001 can all be used as units of counting. There are [ ] 0.001s in 0.01, and there are [ ] 0.01s in 0.1.

4 Fill in the answers.

(a) A decimal number consists of two parts, the _____ part and the _____ part.

(b) 0.67 consists of [ ] 0.1s and [ ] 0.01s.

(c) Two 0.1s are [ ]. Five 0.01s are [ ].

One hundred and twenty-eight 0.001s are [ ].

(d) There are three _____, nine _____,

one _____, two _____ and

nine _____ in 39.129.

(e) In 7.15, the 7 is in the _____ place, representing

seven _____. The 1 is in the _____ place,

representing _____ tenth(s). The 5 is in the _____

place, representing five _____ .

5 Multiple choice questions. (For each question, choose the correct answer and write the letter in the box.)

(a) A fraction whose denominator is 10 can be written as a decimal

with 1 decimal place. $\frac{5}{10}$ is 0.5, or five ☐ .

A. 0.1s   **B.** 0.01s   **C.** 0.001s   **D.** 10s

(b) A fraction with a denominator of 100 can be written as a
decimal with 2 decimal places. $\frac{5}{100}$ is 0.05, or five ☐ .

A. 0.1s   **B.** 0.01s   **C.** 0.001s   **D.** 10s

(c) A fraction with 1000 as the denominator can be written as a
decimal with 3 decimal places. $\frac{5}{1000}$ is 0.005, or five ☐ .

A. 0.1s   **B.** 0.01s   **C.** 0.001s   **D.** 10s

(d) 0.8 equals ☐ .

     A. 0.8 thousandths   **B.** 8 thousandths

     C. 80 thousandths   **D.** 80 hundredths

Challenge and extension questions

6 Write the correct decimal numbers in the boxes.

(a) The number consisting of 65 thousandths is ☐ .

(b) In a decimal number, its hundreds place digit is 6, its ones place digit is 1, its hundredths place digit is 3 and the rest of the digits are all zeros. This number is ☐ .

(c) A decimal number consists of 4 thousandths and 40 tens. It is ☐ .

7 Fiona's maths test score is a bit more than 93 marks but less than 94. What could her maths test score be?

6.4 Understanding decimals (3)

Learning objective Recognise and write decimal equivalents to tenths, hundredths and thousandths

Basic questions

1 Fill in the boxes.

(a) 2 thousands, 4 tens and 1 tenth make the decimal ⬚.

(b) A number consisting of 143 hundredths is ⬚.

(c) A number consisting of 15 ones and 45 thousandths is ⬚.

> A decimal number whose whole number part is not zero is called a mixed decimal.

> A decimal number whose whole number part is zero is called a pure decimal.

(d) Of the numbers 0.41, 5.11, 3.03, 0.8 and 1, the mixed decimals are ⬚ and the pure decimals are ⬚.

(e) Use 2, 1, 0 and a decimal point to make pure decimals with 2 decimal places. They are ⬚.

2 Fill in the boxes. The first one has been done for you.

(a) $0.16 = 1 \times 0.1 + 6 \times 0.01$

(b) $0.448 = \boxed{} \times 0.1 + \boxed{} \times 0.01 + \boxed{} \times 0.001$

(c) $82.57 = \boxed{} \times 10 + \boxed{} \times 1 + \boxed{} \times 0.1 + \boxed{} \times 0.01$

(d) $0.92 = 9 \times \boxed{} + 2 \times \boxed{}$

3 True or false? (Put a ✓ for true and a ✗ for false in each box.)

(a) All decimal numbers are less than 1. . . . . □

(b) There are thirteen 0.1s in $\frac{13}{10}$; it can be written as $\frac{13}{10}$ = 1.3. . . . □

(c) $\frac{39}{1000}$ = 0.390 . . . . □

(d) In a decimal number, the second place to the left of the decimal point is the tens place, while the second place to the right of the decimal point is the tenths place. . . . . □

(e) 0.99 is the greatest pure decimal number with 2 decimal places. . . . . □

(f) 9.99 is the greatest decimal number with 2 decimal places. . . . . □

(g) Mixed decimal numbers are always greater than pure decimal numbers. . . . . □

Challenge and extension question

4 Fill in the boxes.

(a) The number that consists of 6 hundreds, 3 tens, 9 ones, 1 tenth and 8 hundredths is ⬚ .

(b) The quotient of two numbers is 2. If the dividend is multiplied by 100 and the divisor remains unchanged, then the new quotient is ⬚ .

(c) Using four digits 1, 2, 6 and 0 and a decimal point to form

decimals, the smallest pure decimal is [   ] , the greatest

pure decimal is [   ] and the smallest decimal with

2 decimal places is [   ] .

(d) 0.1 is [   ] times 0.001, while 0.001 is [   ] of 0.01.

(e) The greatest decimal with 1 decimal place that is less than 1

is [   ] . The smallest decimal with 2 decimal places that is

greater than 1 is [   ] .

(f) In a decimal, the place value of the digit in the tenths place

is [   ] times the place value of the same digit in the

thousandths place.

6.5 Understanding decimals (4)

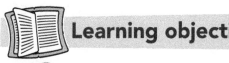

Learning objective Recognise and write decimal equivalents to tenths, hundredths and thousandths

Basic questions

1 Read and write the numbers.

(a) 10.79

Read as: _____.

(b) 22.023

Read as: _____.

(c) 9.304

Read as: _____.

(d) 0.0101

Read as: _____.

(e) 14.90

Read as: _____.

(f) 300.303

Read as: _____.

(g) zero point one seven

Written as: _____.

(h) sixty point nine eight

Written as: _____.

(i)   twenty point zero zero two

Written as: _____.

(j)   one hundred point three seven five

Written as: _____.

(k)   zero point eight zero six zero

Written as: _____.

(l)   one hundred point nine zero zero

Written as: _____.

2 **Write the decimals in the circles as indicated.**
0.9, 54.32, 12.976, 0.31, 46.73, 1.244, 0.18,
9.45, 24.8, 5.77, 7.201, 80.9, 0.07, 100.9.

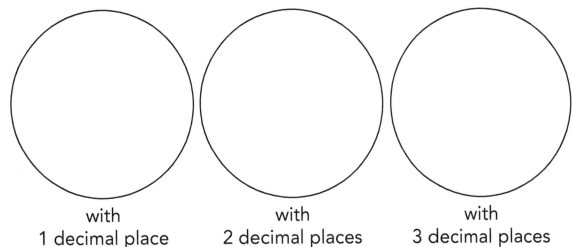

<table>
<tr><td>with
1 decimal place</td><td>with
2 decimal places</td><td>with
3 decimal places</td></tr>
</table>

3 Complete the facts.

(a) There are ☐ pure decimal numbers with 1 decimal place. The smallest one is ☐ and the greatest one is ☐ .

(b) There are ☐ pure decimal numbers with 2 decimal places. The smallest one is ☐ and the greatest one is ☐ .

(c) There are ☐ pure decimal numbers with 3 decimal places. The smallest one is ☐ and the greatest one is ☐ .

(d) The decimal numbers with 2 decimal places that come before and after 0.95 are ☐ and ☐ .

(e) 40.04 is a decimal with ☐ decimal places. The digit 4 to the right of the decimal point is in the _____ place. It means _____ . The digit 4 to the left of the decimal point is in the _____ place. It means _____ . The place value of the digit 4 on the left is ☐ times that of the digit 4 on the right.

4 Use your knowledge to describe decimals.

For example, you can describe 6.78 as follows:

6.78 is a decimal with 2 decimal places.

6.78 is a mixed decimal.

6.78 consists of 678 hundredths.

6.78 is made of 6 ones, 7 tenths and 8 hundredths.

$6.78 = 6 \times 1 + 7 \times 0.1 + 8 \times 0.01$

Now try these on your own.

(a) 13.5

(b) 0.578

Challenge and extension questions

5 Read the following carefully.

If the numerator of a fraction is equal to the denominator, it can be written as the whole number 1.

If the numerator of a fraction is less than the denominator, the fraction is called a **proper fraction.**

If the numerator of a fraction is greater than or equal to the denominator, it is called an **improper fraction.**

A proper fraction can be converted to a pure decimal. When a smaller number is divided by a bigger number, the quotient is less than 1, therefore the whole number part of the resulting decimal is zero. Pure decimal numbers are smaller than 1.

An improper fraction can be converted to a mixed decimal. When its numerator is divided by its denominator, the quotient is greater than or equal to 1. Mixed decimal numbers are greater than or equal to 1.

Therefore, a mixed decimal is $\geq 1 >$ any pure decimal. (Note: the sign $\geq$ means 'greater than or equal to'.)

For example: $1.4 = \frac{14}{10} = 1\frac{4}{10}$. It is read as one and four tenths, and considered as: $1 + \frac{4}{10}$ or $1 + 0.4$.

Now try these.

(a) $1.12 =$ ☐ $=$ ☐   (b) $58.33 =$ ☐ $=$ ☐

6 I am thinking of a mixed decimal. Its whole number part is the greatest two-digit number. The digit in the hundredths place is the greatest single digit, and the digit in the tenths place is the smallest single digit.

This mixed decimal is ☐.

6.6 Understanding decimals (5)

Learning objective Use decimals and convert between different units of measure

Basic questions

1 Write a suitable number in each box. (Note: 1 km = 1000 m, 1 m = 100 cm and 1 cm = 10 mm; drawing not to scale).

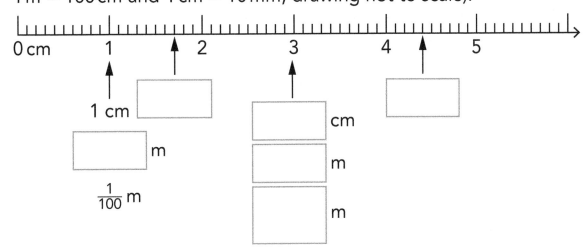

1 cm

[ ]

[ ] m

$\frac{1}{100}$ m

[ ] cm

[ ] m

[ ] m

2 Draw lines to match equivalent measures of length.

$\frac{3}{10}$ m	1 cm	90 mm
10 mm	0.3 m	$\frac{7}{10}$ cm
9 cm	0.7 cm	30 cm
7 mm	0.09 m	0.01 m

3 Use decimals to represent the lengths of the objects below.

(a)

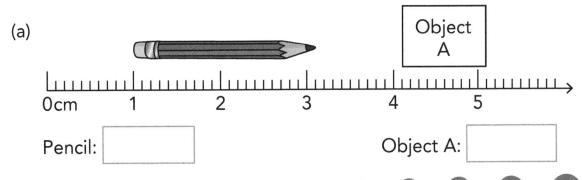

Object A

Pencil: [ ]   Object A: [ ]

(b)

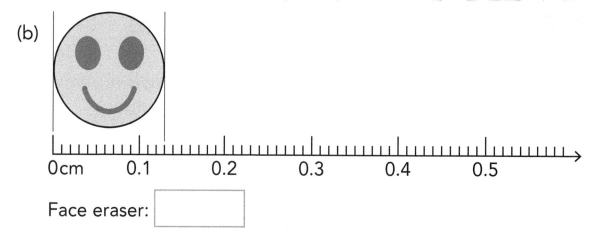

Face eraser: [_____]

4 Use a ruler to measure the length of each side of the triangle and then find the perimeter of the triangle. Measure in centimetres (cm).

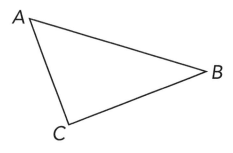

Length of side AB = [_____]

Length of side BC = [_____]

Length of side AC = [_____]

The perimeter of the triangle is: [_____]

Challenge and extension questions

5 Yaseen drank 50 ml more than half of a bottle of milk. There were 530 ml of milk remaining in the bottle. How much milk did the full bottle contain?

6 Fill in the boxes.

(a) 60 cm expressed as a fraction of a metre is ☐ m, and as a decimal is ☐ m. 8 mm expressed as a fraction of a metre is ☐ m, and as a decimal is ☐ m.

(b) A number made up of 3 thousands, 4 ones, 9 hundredths and 8 lots of $\frac{1}{10000}$ is ☐.

(c) There are 10 mangoes in a bag. If 11 people ate 0.1 bags of mangoes, then they ate ☐ bags of mangoes altogether.

(d) There are 100 sweets in a jar. They are divided equally into 100 parts. 25 sweets is ☐ of the whole pack. (Express the answer as a decimal.)

6.7 Understanding decimals (6)

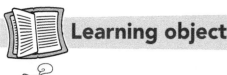

Learning objective Recognise and write decimal equivalents of tenths, hundredths and thousandths

Basic questions

1 Fill in the answers.

(a) There are ⬚ lots of $\frac{1}{100}$ in 0.08.

There are ⬚ thousandths in 0.721.

(b) A decimal number with 3 decimal places has the digit 2 in all the tens, tenths and hundredths places. The rest of the places are zeros.

This number is ⬚.

(c) A number consists of 2 hundreds, 15 ones and 95 hundredths.

This number is ⬚.

(d) A number consists of 1 thousand, 2 tens, 6 hundredths and 1 thousandth.

This number is ⬚.

(e) 803 hundredths make ⬚.

(f) The place value of 8 in the far left digit of the number 8.008 is ⬚ times the place value of 8 in the far right digit.

(g) There are ⬚ 0.01s in 27.93.

(h) Four hundred point zero zero four can be written as ⬚.

It is a _____ decimal number.

(i) 0.707 is read as _____ .It is

a _____ decimal number.

(j) There are ☐ decimal numbers with 2 decimal places that are greater than 3.9 but less than 4.0.

2 Multiple choice questions. (For each question, choose the correct answer and write the letter in the box.)

(a) What value does the digit 6 in each number stand for?

7.26 ☐   60.32 ☐   78.006 ☐   0.619 ☐

    A. 6 tens   **B.** 6 one-tenths

    C. 6 one-hundredths   **D.** 6 one-thousandths

(b) Sixty 0.01s are ☐.

    A. 60   **B.** 6   **C.** 0.6   **D.** 0.06

(c) Five 0.1s are equivalent to ☐ 0.001s.

    A. 50   **B.** 500   **C.** 5000   **D.** 50 000

(d) 7 tens and 7 tenths make ☐.

    A. 700.70   **B.** 7.70   **C.** 70.70   **D.** 70.070

(e) 0.0132 is ☐ 13.2.

    A. one tenth of   **B.** one hundredth of

    C. one thousandth of   **D.** 1000 times

Challenge and extension question

3 A mixed decimal number with 3 decimal places has the following features:

(i) The whole number part is 3 less than the greatest 2-digit number.

(ii) The digit in the tenths place is 6 greater than the digit in the hundredths place, and their sum is 10.

(iii) The digit in the thousandths place is equal to the digit in the tens place.

What is the number? _____

6.8 Comparing decimals (1)

Learning objective Compare and order decimal numbers

Basic questions

1 Let's have a try.

(a) Locate each number below on the number line and mark a small point to indicate its position.

| 5.8 | 2.1 | 4.5 | 7.7 | 0.4 |

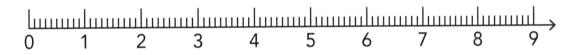

(b) Which of the decimal numbers above is the greatest? ☐

Which one is the smallest? ☐

(c) Is the value of each number related to its distance from the origin, zero, on the number line?

_____

(d) Put the numbers in order, from the greatest to the least.

☐ ☐ ☐ ☐ ☐

2 Fill in the missing words.

When comparing two decimals, we first compare the

_____ part; the greater the whole number part,

the _____ the number. If the whole number parts are

the same, then we compare the digits in the

_____ place; the greater the digit in the

_____ place, the greater the decimal, and so on.

3 Use the method described in Question 2 to help you choose
> or < to write in each ◯.

(a) 1.5 ◯ 1.06

(b) 0.101 ◯ 0.099

(c) 12.25 ◯ 12.26

(d) 0.519 ◯ 0.521

(e) $\frac{1}{2}$ ◯ 0.51

(f) 101.1 ◯ 99.2

4 Put the decimal numbers in order, starting with the greatest.
(a) 0.9, 0.909, 9.09, 0.99

(b) 22.02, 22.20, 22.202, 22.002

5 Study the table below.

Name	Ethan	Marvin	Jaya	Imogen
Height (cm)	1.39	1.33	1.38	1.30
Weight (kg)	25.3	28.2	24	25.2

(a) Who is the tallest? _____

(b) Who is the heaviest? _____

6 In a competition, 3 people were asked to make the same number of model robots. It took Emily 6.5 hours, Joshua 6.25 hours, and Asha 5.9 hours to finish the task. Who worked the fastest?

_____

Challenge and extension questions

7 Write a suitable number in each box.

0.☐6 < 0.07   9.31 > 9.3☐   6.☐4 > 6.54

8 Use the digits 0, 2, 4, 6 and a decimal point to make pure decimal numbers with 2 decimal places. Put them in order, from the least to the greatest.

_____

6.9 Comparing decimals (2)

Learning objective Compare, order and round decimal numbers

Basic questions

1 Compare the numbers and write > or < in each ◯.

(a) 34.1 ◯ 3.41

(b) 0.96 ◯ 0.691

(c) 0.80 ◯ 0.801

(d) 0.103 ◯ 103

(e) $\frac{1}{100}$ ◯ 0.001

(f) 8.139 ◯ 8.130

2 Mark the following decimals on the number line and fill in the boxes.

1.9 4.7 5.2 7.5 9.8 3.4

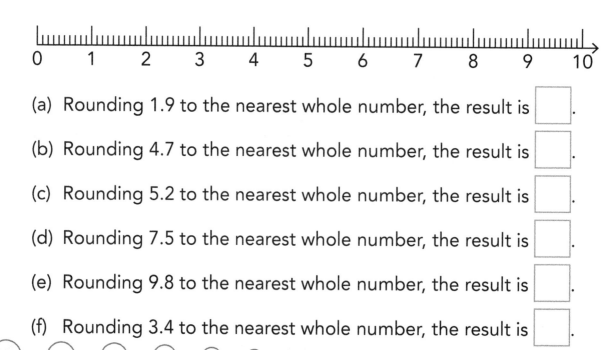

(a) Rounding 1.9 to the nearest whole number, the result is ☐.

(b) Rounding 4.7 to the nearest whole number, the result is ☐.

(c) Rounding 5.2 to the nearest whole number, the result is ☐.

(d) Rounding 7.5 to the nearest whole number, the result is ☐.

(e) Rounding 9.8 to the nearest whole number, the result is ☐.

(f) Rounding 3.4 to the nearest whole number, the result is ☐.

(g) When rounding a decimal with 1 decimal place to the nearest whole number, we look at the digit in the tenths place. If the digit is greater than or equal to ☐, we round it up and the result is the whole number part plus 1. If the digit is less than ☐, the result is just the whole number part itself.

3 Draw lines to match each decimal number in the first row with its nearest whole number.

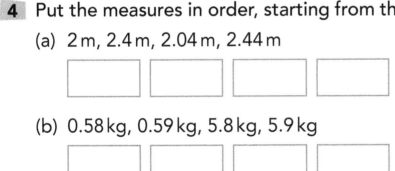

| 1.6 | 212.9 | 1.9 | 213.1 | 0.3 | 2.5 |

| 3 | 2 | 16 | 213 | 212 | 0 | 214 | 1 |

4 Put the measures in order, starting from the smallest.

(a) 2 m, 2.4 m, 2.04 m, 2.44 m

☐ ☐ ☐ ☐

(b) 0.58 kg, 0.59 kg, 5.8 kg, 5.9 kg

☐ ☐ ☐ ☐

5 Thea has £9 and 80p. Mo has £9.08. Who has more?

Remember: £1 = 100p. _____

Challenge and extension questions

6 Round the decimals with 2 decimal places to the nearest whole number.

11.09, 34.77, 15.25, 7.45, 111.118, 1000.41

---

7 The weights of four people, A, B, C and D, are 21.46 kg, 21.52 kg, 21.38 kg and 21.5 kg. They are not in order.

Person A is heavier than Person D but lighter than Person C.

Person D is lighter than Person B, and Person A is heavier than Person B.

What is the weight of each person?

A [　　　] kg　B [　　　] kg　C [　　　] kg　D [　　　] kg

6.10 Properties of decimals

Learning objective Identify properties of decimals, including the value of any zeros

Basic questions

1 Draw three lines of 0.1 m, 0.10 m, and 0.100 m. What do you find?

(a) Write >, < or = in each ◯.

From 1 m = 100 cm, we can see 0.1 m ◯ 0.10 m ◯ 0.100 m.

(b) Fill in the answers.
Properties of decimals: When zeros are added or removed at

the _____ of the decimal part of a decimal number, the value

of the number remains _____ .

2 Write each number in the correct circle.
3.90, 10.005, 300.00, 0.103, 100, 20.002, 1.400

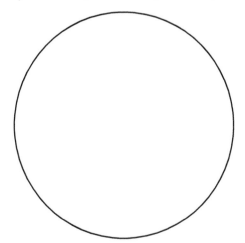

Zeros can be dropped off at the end of the decimal part from any place without changing the value.

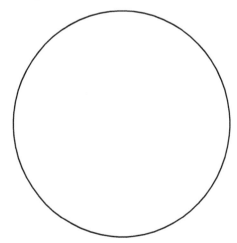

No zeros can be dropped off without changing the value.

3 Use the properties of decimals to simplify the following decimals.

(a) 600.060 = [ ]

(b) 3.500 = [ ]

(c) 700.00 = [ ]

(d) 13.90 = [ ]

(e) 303.330 = [ ]

(f) 10.100 = [ ]

4 Rewrite the decimal numbers as decimals with 3 decimal places without changing the value of any of the numbers.

(a) 1.4 = [ ]

(b) 5.04 = [ ]

(c) 8 = [ ]

(d) 30.400 00 = [ ]

(e) 9.4 = [ ]

(f) 10 = [ ]

5 Compare the numbers and write >, < or = in each ( ).

(a) 1.01 ( ) 1.10

(b) 3.4 ( ) 3.04

(c) 20.2 ( ) 20.200

(d) 4.73 ( ) 4.37

(e) 16 ( ) 16.000

(f) 5.06 ( ) 5.060

(g) 9.89 ( ) 9.98

(h) 13.41 ( ) 13.410

(i) $\frac{1}{100}$ ( ) 0.010

(j) 7.07 ( ) 7.70

6 Multiple choice questions. (For each question, choose the correct answer and write the letter in the box.)

(a) In the following, the decimal number equal to 26.06 is ☐.

    A. 26.60   **B.** 2.6060   **C.** 26.060   **D.** 26.006

(b) The value of a decimal number remains unchanged when zeros are added or removed ☐.

    A. at the end of the number

    B. in the middle of the number

    C. at the end of the decimal part of the number

    D. after the decimal point

(c) There are ☐ numbers greater than 1.1 but less than 1.2.

    A. 0   **B.** 1   **C.** 10   **D.** infinitely many

(d) Of the following numbers, the number whose value is unchanged after each 0 in it is removed is ☐.

    A. 40.34   **B.** 5.910   **C.** 9.02   **D.** 700

(e) Among the following pairs of numbers, the two equal numbers are ☐.

    A. 57.00 and 75.00   **B.** 8.04 and 80.4

    C. 909.90 and 909.9   **D.** 10.01 and 10.10

Challenge and extension questions

7 Fill in the table.

	Rewrite as a decimal with 1 decimal place	Rewrite as a decimal with 2 decimal places	Rewrite as a decimal with 3 decimal places
0.60			
37			
19.900			

8 Put the numbers 0.112, 0.120, 1.1, 0.1021 and $\frac{1}{1000}$ in order from the least to the greatest.

9 A, B, C and D are 4 pure decimal numbers. B has 3 decimal places and its value is between that of A and C. C is 0.01 greater than A and D is 0.001 greater than B but less than C. A is 0.02. What number does C stand for? What numbers do B and D stand for?

C [   ]   B [   ]   D [   ]

Chapter 6 test

1 Fill in the answers.

(a) When rewriting $\frac{29}{100}$ as a decimal, it is [ ]. It is read

as _____. It is a decimal with [ ] decimal places.

It is a _____ (choose 'pure' or 'mixed') decimal. It consists

of [ ] hundredths. Adding [ ] to it makes 1.

(b) The hundreds place is the _____ place to the left of the

decimal point. The hundredths place is the

_____ place to the right of the decimal point.

(c) 5 tenths and 7 hundredths make [ ].

(d) 9.06 plus [ ] hundredths is 10.

(e) 0.37 = [ ] × 1 + [ ] × 0.1 + [ ] × 0.01

(f) 1200 thousandths is [ ]. It is a _____ (choose 'pure' or

'mixed') decimal, consisting of [ ] one(s) and [ ] tenths.

(g) In 57.067, the 7 in the whole number part is in the _____

place and its place value is 7 _____ . The 7 in the decimal

part is in the _____ place and its place value

is 7 _____ .

(h) In 12.21, the value of the digit 1 on the left is [ ] times the value of the digit 1 on the right. The value of 2 on the right is [ ] times the value of 2 on the left.

(i) The whole number part in a decimal is the smallest 2-digit number and the decimal part is the greatest pure decimal with 2 decimal places. This decimal number is [ ].

(j) Two hundred point zero two can be written in numerals as [ ]. Its decimal part has two _____.

Without changing its value, it can be written as a decimal with 3 decimal places, that is [ ].

2 Write the following fractions as decimal numbers.

(a) $\frac{1}{4}$ = [ ]　　(b) $\frac{1}{2}$ = [ ]　　(c) $\frac{3}{4}$ = [ ]

(d) $\frac{1}{10}$ = [ ]　　(e) $\frac{17}{100}$ = [ ]　　(f) $\frac{999}{1000}$ = [ ]

3 Write the following decimals as fractions.

(a) 0.3 = [ ]　　(b) 0.25 = [ ]　　(c) 0.07 = [ ]

(d) 0.21 = [ ]　　(e) 0.75 = [ ]　　(f) 0.191 = [ ]

4 Use the properties of decimals to complete the table. Two have been done for you.

Decimal numbers	Can some zeros be dropped off without changing the value? (Yes or No)	If the answer is yes, write the number after dropping the zeros
0.110	Yes	0. 11
0.205	No	Not applicable (or N/A)
0.7040		
7.000		
68.0100		
200.060		
0.007		
230.0900		

5 Multiple choice questions.

(a) There are ☐ thousandths in 1.2.

   A. 20   **B.** 200   **C.** 120   **D.** 1200

(b) There are ☐ decimal numbers with 1 decimal place greater than 9 but less than 10.

   A. 8   **B.** 9   **C.** 10   **D.** infinite

(c) In a race, it took Freya 2.91 minutes, Jaz 3.1 minutes, Noah 3.05 minutes and Alvin 2.90 minutes to complete the whole course. The person who gained second place was ☐.

   A. Freya   **B.** Jaz   **C.** Noah   **D.** Alvin

(d) The greatest pure decimal number with 2 decimal places is ☐.

   A. 99.99   **B.** 9.99   **C.** 0.09   **D.** 0.99

6 True or false? (Put a ✓ for true and a ✗ for false in each box.)

(a) Inserting or dropping any 'zeros' after the
decimal point of a decimal number will
not change the value of the decimal number. . . . . ☐

(b) After simplifying 0.500 by dropping the
zeros at its end, the number is 5. . . . . ☐

(c) After dropping the last two zeros in 200,
the value remains unchanged. ☐

(d) 0.7 metres and 0.70 metres refer to the same length. . . . . . . ☐

(e) Rewriting 2.4 as a decimal number with 3 decimal places
without changing the value gives 2.004. . . . . ☐

7 Write decimal numbers with 2 decimal places in each box.

(a) 36 pence = ☐ pounds

(b) 4 pence = ☐ pounds

(c) 110 pounds = ☐ pounds

(d) 1 pound 60 pence = ☐ pounds

(e) 3 pounds and 4 pence = ☐ pounds

(f) 90 pence = ☐ pounds

8 Comparing decimals.

(a) Put 0.5, 0.055, 0.505 and 0.550 in order from the greatest to
the least.

☐ > ☐ > ☐ > ☐

(b) Put the following measures in order from the shortest to the
longest: 5 km 4 m, 0.0054 km, 5.04 km, 5.40 km.

☐ < ☐ < ☐ < ☐

9 Round the following decimals with 1 decimal place to their nearest whole numbers. Write the results in the boxes.

15.9	0.1	10.4	119.5

10 Use the digits 0, 1, 3, 5 and a decimal point to write numbers as indicated.

(a) All decimal numbers less than 1 and with 3 decimal places.

_____

(b) All decimal numbers greater than 5 and with 3 decimal places.

_____

(c) All decimal numbers that contain zero, but the zero is not read out, and have 2 decimal places.

_____

11 In a maths test, Edward's score is 92.35 marks. Theo's score is 2.5 marks higher than Edward's. Lily's score is 0.7 marks lower than Edward's. Joe's score is 3.4 marks higher than Edward's, and May's score is 0.5 marks lower than Edward's. Among the five pupils, who has the highest score?

_____

Notes

Notes

Notes